Mobile Technologies

Routledge Research in Cultural and Media Studies

**19. The Contemporary Comic
Book Superhero**
Angela Ndalianis

20. Mobile Technologies
From Telecommunications to Media
Edited by Gerard Goggin &
Larissa Hjorth

Mobile Technologies

From Telecommunications to Media

Edited by
Gerard Goggin and
Larissa Hjorth

Routledge
Taylor & Francis Group
New York   London

First published 2009
by Routledge
270 Madison Ave, New York, NY 10016

Simultaneously published in the UK
by Routledge
2 Park Square, Milton Park, Abingdon, Oxon OX14 4RN

Routledge is an imprint of the Taylor & Francis Group, an informa business

Transferred to igital Printing 2009

© 2009 Taylor & Francis

Typeset in Sabon by IBT Global.

Library of Congress Cataloging in Publication Data

Mobile technologies : from telecommunications to media / edited by Gerard Goggin &
    Larissa Hjorth.
    p. cm. — (Routledge Research in cultural and media studies ; 20)
    Includes bibliographical references and index.
    1. Cellular telephones—Social aspects.   2. Interpersonal communication—
Technological innovations—Social aspects.   3. Social interaction—Technological
innovations.   4. Communication and culture—Technological innovations.
I. Goggin, Gerard, 1964–   II. Hjorth, Larissa.
    HE9713.M663 2009
    303.48'33—dc22
    2008030161

ISBN10: 0-415-98986-8 (hbk)
ISBN10: 0-415-87843-8 (pbk)
ISBN10: 0-203-88431-0 (ebk)

ISBN13: 978-0-415-98986-2 (hbk)
ISBN13: 978-0-415-87843-2 (pbk)
ISBN13: 978-0-203-88431-7 (ebk)

Contents

Figures

Acknowledgments

Many of the papers in this volume had their beginnings in papers delivered at the international conference on Mobile Media, held at the University of Sydney, 1–3 July. The proceedings of the conference forms a companion volume to this book. Our special thanks go to Genevieve Bell and Intel Corporation for sponsorship that made this conference possible. We also gratefully acknowledge the support of the School of Letters, Arts and Media and the Faculty of Arts, University of Sydney. We also would like to thank Ryan Sengara for his wonderful work in helping organize the conference.

Gerard's work on this collection was supported by a ARC Australian Research Fellowship project Mobile Culture: A Biography of the Mobile Phone (DP 0453023). He wishes to acknowledge the support of the Journalism and Media Research Centre, University of New South Wales. Gerard also thanks the Centre d'Estudis Australians, Universidad de Barcelona, and especially Sue Ballyn, for a visiting professorship, which made it possible for him to pursue important parts of the editing and research for the book. Finally, Gerard wishes to thank his partner Jacqueline and children Liam and Bianca for their unstinting support, and continued interest in things mobile, telephonic and mediated.

Larissa would like to thank CICC (Communication Interface and Culture Content) Research Group at Yonsei University (Seoul, South Korea) for the research fellowship in 2007 and RMIT University Games Programs.

Gerard Goggin
Journalism and Media Research Centre
University of New South Wales, Sydney

Larissa Hjorth
Games Programs
Royal Melbourne Institute of Technology
June 2008

Part I
Reprising Mobile Theory

1   The Question of Mobile Media

Gerard Goggin and Larissa Hjorth

Mobile phones are now well advanced in their global diffusion. Many aspects of mobile telephony are firmly and unremarkably ensconced in everyday life. Many other facets of the mobile phone still attract much public fascination. One of these is the phenomenon of mobiles *becoming* media.

The idea of mobiles as media is no idle conceit, and it can be substantiated in a number of senses. Cellular mobile phones, devices, and networks can still be usefully conceptualized as telecommunications. And although telecommunications has been a minor and often overlooked part of media systems, it is actually vitally important. The mobile phone quite obviously presented itself as something that followed in the trajectory of the telephone, and then telecommunications. Quickly, though, mobiles have amounted to something quite different. As the substantial scholarship on text messaging shows, this extension of portable mobile phones built upon predecessor messaging technologies (notably the pager and wireless telegraphy, but also the fixed-network technology of the telex). Text messaging took on a life of its own—spawning a career in technology that quickly moved beyond mere signal or data to direct suturing into youth culture, interactive television, mobile commerce, and so on.

What text messaging underscores is that the consumption of mobile phones is an important part of contemporary media around the world, in ways that we still only dimly understand. The importance of mobiles to media is especially pronounced not only in many Western countries but also in the new geopolitical forces represented by countries such as India and China, as well as significant groups in developing countries. For instance, the mobile has been vital for conveying news, whether via voice calls or text messaging, in various countries where political conditions have been inimical to free and diverse press, and where mainstream media, or even new Internet-based media, such as Web sites, lists, or blogs, have been subject to censorship and repression. Hence, the use of the mobile for reporting and circulating news in countries such as Burma, the much-publicized role the mobile plays in activism and organizing in many countries, or the inventive way that satire and jokes with clear social and political valence have been circulated via mobiles in a number of countries.

What is now also evident is the incorporation of mobiles into mainstream, minority, alternative and citizen's media alike as a vital channel in a convergent, cross-platform approach to media. Obvious examples are the use of mobiles to provide interactivity and allied platforms and capacity to refresh and extend television—evident in the much-publicized use of mobiles in international-format programs such as *Idol, Big Brother*, and many others. With mobiles in the pocket of the citizen and denizen, and key to new architectures of popular culture, television producers and broadcasters internationally have responded accordingly. The rise of user-created content (UCC) has become synonymous with the deployment of mobile media as a foray into Web 2.0 and social networking systems (SNS). The case of news, already introduced, is also a good case in point. With mobiles come the affordances of alerts, mobile Internet, and the ability to watch and forward video, among other attributes. Unsurprisingly, many mainstream news outlets, especially in their online forms, offer a range of news especially repurposed, customized, or designed for mobiles.

An earlier and much more pervasive example is the case of mobile music. Music is another area of media, often overlooked, but vital to culture generally. Popular music has been avidly consumed by users through mobiles, with the invention of ring-tones, but also through accompanying micropayment systems (notably pioneered in Japan with the i-Mode system). Mobile technologies are now intersecting and interacting with other recent developments in music, including online music, digital formats, and portable music players. With their increased storage capabilities, and the possibilities for exchanging music either freely (via protocols such as Bluetooth) or via mobiles or Internet, mobiles have a dual function as digital music organizer, players, and sharing device—thus figuring in the media and music ecologies inhabited by technologies such as the MP3 player, iPod, or peer-to-peer file sharing.

Mobiles, then, are a strategically important site of innovation, change, and re-invention of older, existing media. However, mobiles are being re-imagined as media in far bolder guises still, with the advent of forms such as mobile books, mobile television, mobile Internet, and mobile games.

AFTER TELECOMMUNICATIONS

In the growing field of scholarship and critical examination of mobile technologies, then, this collection focuses on the transition from telecommunications to media. The warrant for the collection lies in the distinct sense we have of much public fascination with the mobile's media turn, many design, corporate, and service offerings that position the cell phone front and centre in media spaces, and considerable consumer and user response, though often from unexpected angles. We have keenly felt the lack of a resource that brings together critical accounts, case studies, and analyses of what is unfolding here. There are other treatments that address facets of the topic—such as the useful 2005 collection *Mobile Media: Content and Services for Wireless Communications*[1]—and also works that investigate what the trajectories come after the classic period of the cell phone.[2] However, our focus is neither solely on the framing of this new epoch, and its forms, as "mobile content," nor as primarily about "mobile communication." Rather, we are interested in focusing upon, and promoting inquiry into, the changes in mobiles that the shift into media brings; and the related changes in the nature of contemporary media that mobiles catalyze.

To approach the question of the media transformation of mobiles, the volume is organized into five parts. The first part, "Reprising Mobile Theory," offers two important new papers on key topics that provide a bridge from current mobile studies. In their "Intimate Connections: The Impact of the Mobile Phone on Work/Life Boundaries," Judy Wajcman, Michael Bittman, and Jude Brown provide a fresh perspective on an recurrent and still vitally important theme in the social shaping of mobile technology. This is followed by a contribution from the profound and ceaseless theorist of mobiles, Leopoldina Fortunati, inquiring into the concept of gender. The second part, "Youth, Families, and the Politics of Generations," also deals with a keenly studied and debated area of mobiles—taking stock of where the mobile and its coproduction of the social is at, but also gauging the changes evolving with new patterns of consumption and user innovation, interacting with the affordances of mobile media devices. Leslie Haddon and Jane Vincent's "Children's Broadening Use of Mobile Phones" acquaints us with adroitly conducted, closely observed, and carefully drawn research about the new things that British children are doing with mobiles, and what this means. In a vintage paper, Rich Ling, another leading figure in mobile studies, thoughtfully and tellingly takes up the theme of teenagers, and looks at what we now know about how mobiles fit into the process, politics, and temporality of becoming (and unbecoming) a teen. Misa Matsuda's insightful "Mobile Media and the Transformation of Family" delves behind the "aura-of-crime" phenomenon to expose how this is orchestrated within parental and sibling micropolitics—such technics are particularly apparent in children's deploying the *keitai* (Japanese abbreviation for mobile phone) as "mom in the pocket." In their "*Purikura* as a Social Management Tool," Daisuke Okabe, Mizuko Ito, Aico Shimizu, and Jan Chipchase start with an established media form, the Japanese *purikura*, or photo sticker booth or collecting book. They then give us a fascinating study in how the *keitai* becomes part of this wider media ecology.

From the reconsideration of well-established themes in mobile studies and the review and recasting of these in grappling with mobile's posttelecommunications environment, we move to two sets of detailed studies that explicitly take up the problematics of what are mobile media and how might we define them (Part III, "Mobiles in the Field of Media"), and the concomitant issues of what are their relationships and position in regard to predecessor and continuing media forms (Part IV, "Renewing Media Forms").

Opening Part III, Jonathan Donner extends his invaluable work in his "Mobile Media on Low-Cost Handsets: The Resiliency of Text Messaging among Small Enterprises in India (and Beyond)"—making an important argument for the everyday yet powerful framing of mobiles by users in developing countries. The original and noted thinker on technology, Harmeet Sawhney, contributes an important paper that disentangles the entwining of the two great contemporary media in his "Innovations at the Edge: The Impact of Mobile Technologies on the Character of the Internet." Virpi Oksman analyzes the interplay of multimedia in her "Media Contents in Mobiles: Comparing Video, Audio, and Text." Stuart Cunningham and Jason Potts's "New Economics for the New Media" draws on groundbreaking work in evolutionary economics in an effort to rethink the markets, industries, structure, and relations of creativity, consumption, and production, and concepts of the social, in which mobile media are set. Larissa Hjorth draws upon the domestication and new media traditions, as a fertile and revealing frame for situating mobile media, in her paper "Domesticating New Media: A Discussion on Locating Mobile Media." In Part IV, we commence with a fascinating, historically grounded study of mobile television in Italy, with Gabriele Balbi and Benedetta Prario's "Back to the Future: The Past and Present of Mobile TV." In their "Net_Dérive: Conceiving and Producing a Locative Media Artwork," Atau Tanaka and Petra Gemeinboeck offer a rich provocation and model for thinking about advances in music, location technologies, and art as they meld as mobile media. While there is a rich literature on Internet and online news, this staple of media has not been given the attention it deserves in relation to mobiles, and this is redressed by Liu Cheng and Axel Bruns's "Mobile News in Chinese Newspaper Groups: A Case Study of Yunnan Daily Press Group." This paper is nicely complemented by Wendy Van den Broeck, Bram Lievens, and Jo Pierson's discussion of electronic reader and paper technologies and nascent reader responses and practices in their "Re-inventing Newspapers in a Digital Era: The Mobile E-Paper."

The concluding part of the book is titled "Mobile Imaginings," as the four papers to be found there raise an ensemble of cross-cutting issues and try to trace and speculate upon their deep cultural significances. Kathy Cleland's "Face to Face: Avatars and Mobile Identities" is one of the first accounts we know of that seek to theorize the abiding issue of avatars for mobile platforms. Dong-Hoo Lee's marvelous "Re-imagining Urban Space: Mobility, Connectivity, and a Sense of Place" discusses space and place through the geomorphologies, experiences, and relations of mobile social media, such as Cyworld. Kate Crawford contributes a pioneering essay on the microblogging of the Twitter technology, and its poetics, in her "These Foolish Things: On Intimacy and Insignificance in Mobile Media." Finally, Nicola Green looks at the implications of new mediated mobilities for the much-discussed and strategically important thematics of memory in her chapter "Mobility, Memory and Identity."

CHALLENGES FOR FRAMING MOBILES AS MEDIA

As these contributions underline, when mobile phone transforms into mobile media there are resistances, contestations, and challenges that arise. In the epoch of user-created content and Web 2.0 in which social labor and capital become the precepts for global media vernaculars, it is through the rubric of mobile media that one can begin to understand these new forms of material and immaterial labor and attendant intimacies. One of the first challenges is to recast ways of conceptualizing forms of communication and expression across both material and symbolic dimensions. It is important not to see "mobile media" as "new" but rather a recontextualization of older media, ideologies, and practices.

Another challenge is to rethink what constitutes creativity—such as the distinctions between amateur and professional categories—and what that entails within the context of the "Swiss army knife" of mobile media. On the one hand, mobile media has symbolized a democratization of media, affording the everyday user with the ability to become a photojournalist or micromovie filmmaker. Within this line of thinking, we can see new forms of creativity and artistic labor evolving—akin to the webcam revolution. The particular characteristics of mobile media—portability, ubiquity, miniaturization, and personal media, to name but a few—provide new forms of visuality, textuality, aurality, and haptic vernaculars. These emerging aesthetic trajectories offer a lens for rethinking media practice in which new forms of Web 2.0 distribution, such as YouTube, create new contexts and modes of audience participation. On the other hand, it is important to look beyond the smoke-and-mirrors rhetoric of the "produser" paradigm and see the persistence, and even amplifications, of inequalities across various social groups. In particular, it is important to recognize that much of the emerging forms of mobile media creativity are coming from developed contexts.

Thirdly, it is pivotal to recognize the dialectic between users and nonusers, and how this dynamic informs the scope and discourse of mobile media today. It is easy to look towards the early-adopters—to fetishize the technological savvy—without acknowledging the role they play as part of broader social, cultural, and economic shifts. Moreover, within the user and non-user paradigm lie some broader issues about the emerging forms of mobility and immobility that greet the twenty-first century. These mobilities—and immobilities—can take various forms such as capital, people, ideas, and objects. Through the lens of mobile media, as a global phenomenon that is marked by localization, we can begin to reconceptualize the symbolic and material ways in which inequalities are played out within postmodernity.

Fourthly, as mobile phones depart from the rubric of telecommunication and into the realm of mobile media, so too do we need new models for equating and analyzing the phenomenon. Whilst mobile phones have long attracted—and necessitated—interdisciplinary approaches drawing from media and communication, sociology, cultural studies, and anthropology, it is the migration into mobile media that urges us to consider more sophisticated and complex apparatus for analysis. This is when such traditions as new media can provide insight into the convergence and divergence of old and new media in the form of mobile media. So too mobile media can offer innovative ways to reconfigure these disciplines as well as a reconceptualization of interdisciplinary practice within the twenty-first century.

Often, youth has been conflated with the prerogative of being the subjects for new technologies. However, as mobile phones unevenly transform into mobile media, issues about demographies are contested. Just as the rise of mobile phones was one marked by surprising adaptations—such as the phenomenal spread and impact of Short Message Service (SMS) to the localized forms of digital storytelling and emerging new creative forms of expression in the form of camera phone imagery—so too the nascent rise of mobile media must be conceptualized in terms of localized contingencies, disruptions, connections, and disconnection. Thus, the challenges for research agendas for mobile media are the need to grapple with both new and revised media practices, labor, and politics. In order to do so, more cross-generational, transnational, and longitudinal studies need to be conducted to further contextualize the phenomenon for its full complexities.

NOTES

1. *Mobile Media: Content and Services for Wireless Communications*, edited by J. Groebel, E. M. Noam, and V. Feldman (European Institute for the Media, 2005).
2. For instance, see *After the Mobile Phone? Social Changes and the Development of Mobile Communication*, edited by Maren Hartmann, Patrick Rössler, and Joachim Höflich (Berlin: Frank & Timme, 2008).

2 Intimate Connections
The Impact of the Mobile Phone on Work/Life Boundaries

Judy Wajcman, Michael Bittman, and Jude Brown

If there is one single artifact that characterizes modern life in the fast lane it is the mobile phone. Both as a practice and a symbol, it epitomizes what it means to think about mobility and immobility today. As the mobile phone increasingly becomes a platform for mobile media, so too does it become the harbinger for debates around the convergence of work and personal life. Our intention here is to consider the impact of this iconic technology on work/life balance, which over the last decade has become a major area of social science investigation and policy debate.[1] In the United States, for example, numerous studies measure what is referred to as "home-to-job and job-to-home spillover," whereby experiences in one domain moderate the experiences in the other.[2]

Spillover can be positive or negative, but most of the research has been on negative spillover, when demands from the two domains of job and home compete for an individual's time, energy, and attention. In Australia and Britain, organizational, government, and academic discourses favor the gender-neutral language of work/life balance or work/life integration, in contrast to earlier discourses of family-friendly policies.[3] Whatever the language, the shared assumption is that the boundaries that once separated work and home life are increasingly permeable.

To date, little connection has been made between this research and the literature on the mobile phone and information and communication technologies (ICTs) more generally. But this has not prevented mobile wireless technologies from being routinely credited with blurring the historically separate spheres of public and private, as work becomes a task that can be carried out in almost any location and domestic concerns can equally be teleported.[4] While for some commentators these developments represent a threat to the quality of modern life, for others they represent new opportunities for integrating the spheres of work and family.[5] On the one hand, many have claimed that the mobile phone creates the communicative potential of perpetual contact, allowing (perhaps even encouraging) work problems to colonize the social spaces and times once reserved for family life.[6] On the other hand, in facilitating the microcoordination of family arrangements, one might expect that mobile phones play a very positive role in maintaining work/family balance.[7]

In fact, we are only beginning to understand how the rapid diffusion of communication technologies has affected these dynamics, still less whether the practice of using these devices might be reconfiguring home/work boundaries. In this paper, we present some preliminary findings of an Australian survey, purposely designed to study how individuals and households are deploying the mobile phone to integrate the different dimensions of everyday life.

Before that, however, we review current research on the impact of mobile phones on work/life balance. We will argue that much of this material takes for granted the very boundary separation between the workplace and home that it purports to be examining. One of the reasons for this is that the debate on work/life balance and spillover tends to adopt a rather static model of the home as a fixed space in which family life is experienced and lived. Changes wrought by mobile devices provide the opportunity to reexamine this conceptual framework and to consider whether "boundary permeability" is the best way to frame the complex issues associated with the shifting work/family interface. If mobile communications have indeed removed the basis of the older boundary between work and home, then has this boundary disappeared, has it been replaced by other social practices, or has its very character been modified?

BOUNDARIES OF TIME AND SPACE

The context for these debates about work/life boundaries is the traditional gendered divide between market work (male) and domestic work (female). With the increasing integration of women into the workforce, and the predominance of dual-earner families, the spatial, temporal, and gender divisions that once characterized the distinct spheres of family and work are no longer such clear demarcations. As a result, a range of discourses referred to earlier have been developed to address the new intersections between employment and home, with key commentators in Australia, Britain, and the United States arguing that the conflict between household and workplace "temporal regimes" is intensifying.[8] Time has become a critical resource as pressures on working time, family time, and individual time challenge people's sovereignty over their own time.

The boundary between work and home life was one of the principal ways that many people controlled their time. Before the emergence of wireless telephony, separate phone lines (with separate numbers) were designated for business and home. Using the fixed telephone at work was seen as a significant tool in helping employees manage family issues while at work, such as making care arrangements, checking up on family members, and being contactable in the event of any problems. Conversely, most employers would limit access to home telephone numbers to "emergency" or crisis situations.

When the Workplace Industrial Relations Survey was carried out in Australia in 1995, it found that a significant proportion of employees did not have access to the phone during work time for personal family calls.[9]

Since then, in Australia and elsewhere, access to mobile phones has become ubiquitous. There are over 18 million mobile phone services in operation in Australia (population 21 million) compared to 11.5 million landline phones.[10] The penetration rate of the mobile phone is estimated to be around 90 percent, with over a quarter of all calls made in Australia done so via mobile phones. Last year Australians sent over eight billion SMS messages (at least 300 messages for each subscriber).[11] These numbers suggest that the mobile phone has become an integral part of everyday life.

Digital communication devices provide people with an opportunity to break from traditional uses of space and time in delineating the realms of work and home.[12] As David Morley expresses it, communications technologies have

> . . . the simultaneous capacity to articulate together that which is separate (to bring the outside world into the home, via television, or to connect family members, via the phone, to friends or relatives elsewhere) but, at the same token, to transgress the (always, of course, potentially sacred) boundary which protects the privacy and solidarity of the home from the flux and threat of the outside world.[13]

Whereas time was largely managed through this boundary separation between the workplace and home, today's technologies create the potential for perpetual contact, making the pace of events instantaneous and simultaneous. So one might expect that mobile phones play a very positive role in maintaining work/family balance. According to Pan-European surveys, most people agree that the mobile phone helps coordinate family and social activities.[14] However, does it necessarily follow that this technical feature of mobiles—that respects no spatial boundaries—will undermine the divisions between work and home?

Not surprisingly, work spilling over into family time has been the subject of several studies from a management and organiational perspective. For example, a recent Canadian study of managers, professional and technical workers, actually defines cell phones, laptops, home computers, BlackBerry devices, and PDAs (personal digital assistants) as work extending technology (WET)—meaning the act of engaging in work-related activities outside of regular offices hours in locations other than the business office.[15] While documenting the dramatic increase in the number of workers using these technologies, their most interesting finding is that the same features of WET that increase perceived control and facilitate communication also appear to be the source of many oppressive features. However, the lack of a clear boundary between work and home resulted for these workers in a sense of having less time available for spending with the family.

Noelle Chesley's research, based on professional and managerial career couples in the United States, found that frequent use of communication technologies is linked to greater work/family boundary permeability, lowering family satisfaction.[16] While work/family spillover was found more typically among men, and family/work spillover among women, it was only women who reported that taking family-related calls at work was stressful. In this way, the author argues that communications technology may be reinforcing gendered work/family boundaries, as family worries and responsibilities appear to be more likely to influence women's outcomes.

PRACTICING THE FAMILY, PRACTICING TECHNOLOGY

While these studies provide valuable information about the relationship between mobile devices and work/family boundaries, their primary focus on managerial and professional employees is not necessarily indicative of more general societal trends in mobile-phone usage. More fundamentally, however, we need to question whether "boundary permeability" is the best way to frame the issues associated with a shifting work/family interface. As people further adopt and incorporate mobile phones into their everyday lives, it may be that the spatial, organizational, and even psychological border between time at home and time at work will lose its salience. This suggests that we should focus less on the impact of these devices on the boundaries themselves and instead reformulate the question in terms of the control that individuals can and do exercise over what passes through these boundaries.

Much of the literature on spillover theory and on work/life balance adopts a rather static model of the family as a bounded microsystem, exchanging with other relatively bounded entities.[17] David Morgan's focus on *family practices* is a useful corrective to spillover approaches, stressing that family life is always continuous with other areas of existence: "family practices are not necessarily practices which take place in time and space conventionally designated to do with 'family,' that is the home."[18] Rather, viewing the family as actively constructed through the day-to-day activities of its members enables us to see paid work outside the home as constituting part of family practices. For Morgan, then, individuals are *doing* family, instead of passively residing within a pregiven structure: family is designated less as a noun and more as a verb.

This identification of the home as the domestic household that is a bounded place has also been interrogated by Mary Douglas.[19] Douglas argues that while the home is located, it is not necessarily fixed in space—rather, home starts by bringing space under control. For Douglas, a home is not only a space but involves regular patterns of activity and structures in time. The solidarity that is the essential foundation of the home is made possible by the highly complex forms of coordination maintained by open, constant communication. The burden of the home's organization is largely carried out by its adherence to conspicuous, fixed times and timetabling for joint and separate activities of its members. (Hence, the emphasis on common presence at fixed times, such as family meals.) If the domestic home is characterized not by "the stoutness of the enclosing walls"[20] but rather by the complexity of coordination, then the mobile phone's capacity for enabling "microcoordination"[21] makes it the ideal tool for *doing* family.

The emphasis on family life as practices resonates with "constructivist" approaches in the social studies of science and technology (STS).[22] Rather than casting ICTs as instruments of communication, the STS literature emphasizes that new technologies reconfigure relationships between people and the spaces they occupy, altering the basis of social interaction. Standard sociological readings of ICTs tend to give primacy to social relations, treating them as existing prior to and outside the intervention of technology. By contrast, approaches more in tune with the social studies of science and technology conceive of the technical as part of the constitution of the "social." In other words, the "social" and the "technical" are not separate spheres, but one and the same. Morgan's conceptual ploy of treating the family as a verb has striking parallels with actor-network theory in which nonhuman objects are conceived of as actants and society is viewed as a *doing* rather than a *being*. The construction of technologies is then a moving relational process achieved in daily social interactions: entities achieve their form as a consequence of their relations with other entities.[23] So STS conceptualizes social interests, and even identities, as being coproduced with artifacts.

The "domestication" framework, in particular, has sensitized researchers to some of the complex processes at work in incorporating ICTs into everyday life.[24] In line with social shaping principles, domestication foregrounds user agency in the way people decide to use and incorporate these artifacts in their lives. It also takes into consideration the impact that context may have on this use, and regards use as collective rather than individual. In conceptualizing technologies as sociomaterial configurations, such analyses are attuned to the capacity of mobile modalities to foster novel patterns of social interaction, altering the quality of intimate relationships. In other words, as with other technological innovations, mobile devices may be ushering in a range of new social practices, communication patterns, and corresponding forms of life. It may thus be time to reformulate our analysis of work/home permeability in terms of how family practices are being mutually configured by mobile communication technologies.

In this vein, Christian Licoppe suggests that ICTs provide a continuous pattern of mediated interactions that combine into "connected relationships," blurring the boundaries between absence and presence.[25] Rather than thinking about tasks and relationships as being located in one physical sphere or the other, he argues that new communication devices (such as the mobile phone) are not just added to other devices or substituted for rival uses. Rather, "it is the entire relational economy that is 'reworked' every time by the redistribution of the technological scene on which interpersonal sociability is played out." Noting the frequency and short duration of mobile-phone calls and mobile text messaging (in France), he argues that communication practices are being redirected towards connected interpersonal communication practices. Importantly, Licoppe stresses that this "connected" mode coexists with previous ways of managing "mediated" relationships, representing the emergence of a new repertoire for managing social relationships.

If connected relationships blur the boundaries between absence and presence, then by implication they are disrupting and rearticulating the spheres of public and private. A key question that arises then is might the constant connection afforded by mobile modalities be changing the quality of intimate relationships? After all, families remain a crucial relational entity playing a fundamental part in the intimate life and connections between individuals. In this vein, Chantal de Gournay argues that we are witnessing a fetishization of communication insofar as the mobile phone is used not for interaction but rather for a "fusional" relationship, so that one takes a part of the other person with one, sure of his or her availability for permanent and total possession.[26] The mobile phone is a bit of their intimacy that people are taking outside the home, "seen almost as part of the body, intended to reassure and compensate for all emotional wants."

However, much of the writing on the impact of telephony has not been sufficiently attuned to gender dimensions. Australia is an exception as, thanks to Ann Moyal, we had one of the earliest studies of the gendering of the landline.[27] Moyal found that masculine and feminine modalities of telephone usage exhibited clear differences, with a feminine culture of " 'kinkeeping,' caring, mutual support, friendship . . . and community activity" playing a central part, very rarely found among men. Lana Rakow's American study of the landline similarly reports that: "telephoning is a form of care-giving . . . gendered work . . . that women do to hold together the fabric of the community, build and maintain relationships."[28] The requirement to perform this affective or emotional work is part of the unequally distributed gender division of labor.

As feminist sociologists of the family have long pointed out, this care work is an aspect of intimacy underemphasized in mainstream writing.[29] Whether the mobile phone will be incorporated into existing patterns of gender differentiation or disrupt gendered usage is still an open question. I have argued elsewhere that there is no reason why the mobile phone should be seen as a gendered artifact, as it does not carry the masculine connotations of, for example, computers that are still identified with hacker culture.[30] This is one of the issues that our study explores.

MOBILE-PHONE STUDY: METHODS

The remainder of this chapter explores these themes in relation to findings from our study of mobile-phone practices in Australia.[31] The study design involved a sample recruited from a panel of online respondents maintained by AC Nielsen. The sample of 1,358 individuals in 845 households was collected from March to May 2007, and comprises all individuals in households aged fifteen years and older. The characteristics of the panel match those of the total population which is online (currently estimated to be approximately 60 percent of Australian households).

The survey consists of three components—a questionnaire, a phone log, and a light time diary. In this chapter, data from the questionnaire and phone log are analyzed. The questionnaire asked respondents about the following areas: ownership and use of mobile phones; the perceived impact of mobile phone use on work and life balance; perceived effects on work and work/family spillover; effects on social support networks; and the phone's role in coordination and control.

The mobile-phone log asked respondents to give details about their ten most recent phone calls and SMS text messages, both made and received, using the information already stored in their handsets. Information was collected on whom the call/text message was to or from (for example, spouse, work colleague, service provider), the gender of the caller, and the date and time of the call. While previous research has utilized billing information, our log method captures the substantial number of prepaid customers for whom no billing records exist, estimated to be around half of the mobile market in Australia.

MOBILE-PHONE STUDY: RESULTS

Almost 90 percent of the sample uses a mobile phone, with two-thirds having owned a phone for more than five years. We began the questionnaire by asking people about their reasons for making calls and sending SMS messages on the mobile phone. Calls on the mobile phone are predominantly made for social or leisure purposes (32 percent) or for managing home and family (29 percent). Other interpersonal contacts account for 15 percent of the reasons for making calls and only 24 percent of calls are related to work or study. There are marked differences between men and women in the purposes for which calls are made. Over a third of men (38 percent) use their mobile phone to make calls for work or study activities, whereas only 11 percent of women use it for this purpose. Social uses of the phone account for the remaining 89 percent of women's calls. Text messages are even more socially oriented and a smaller proportion of both men's (15 percent) and women's (5 percent) texts are devoted to work or study.

An analysis based on information retrieved from the handset log of calls, shown in Figure 2.1, reveals that user's perceptions of their mobile-phone usage are based on a good recall of their actual behavior. Of the 9,714 calls, contacting family (48 percent) and friends (26 percent) is the overwhelming use. Similarly, for both men and women, by far the most common recipients of text messages are family (45 percent) and friends (43 percent).

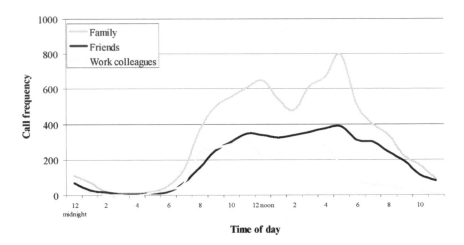

Figure 2.1 Frequency of calls by time of day and call recipient.

Conversely, only a small proportion (16 percent) of calls are work-related. Among calls to family members, for both men and women, the highest proportion is calls and text messages to one's spouse (18 percent). Women are disproportionately likely to phone their children (11 percent), parents (12 percent) and extended family (11 percent). On the other hand, in general, men are more likely to use the mobile for work-related calls, and this holds true even when employment is taken into account. Employed men devote 23 percent of their calls to work-related purposes, while for employed women the percentage is 15 percent.

As expected, the phone log (Figure 2.1) reveals that work-related calls are mostly confined to standard working hours, rising sharply after 8 a.m. and declining steeply around 5 p.m. Family calls are less frequent in the morning than in the afternoon, rising at the time school ends, and having a pronounced peak before the evening meal. Throughout the evening, family calls are at a much higher level than work-related calls. Contacting friends begins mid-morning and remains sustained throughout the afternoon and early evening. During the entire evening, communications with friends are at a higher rate than work-related calls. The heavy use of the mobile in the evening for contacting family and friends (and not job-related tasks) is consistent with our view that the main purpose of the mobile phone is for social contact.

BALANCING LIFE AND WORK

Employed respondents were asked to rate "What impact has the use of your mobile phone had on your ability to balance your work and home/family/

personal life?" Very few respondents report that the mobile phone has a negative impact on their work/life balance (3 percent). A higher proportion of respondents than anticipated (43 percent) say that it has had no effect. Significantly, however, more than half of the respondents believe that the mobile helps them to balance their family and working lives.

A major property of mobile phones is its use for the microcoordination of family arrangements and work schedules. Indeed, Rich Ling has argued that "the mobile telephone has started to change the ways in which we organize and coordinate our everyday lives."[32] The idea of microcoordination involves the "softening of schedules," adding slack to the more inflexible nature of time-based arrangements. The scheduling of events is relaxed through an ongoing sequence of reciprocal phoning ahead that enables meetings to be renegotiated "on the fly" so that the needs of parties can be progressively accommodated.

In order to explore this function, we asked respondents in multiperson households: "How significant are the following reasons ('planning meals'; 'arranging to meet with family/household members'; 'arranging to deliver goods or children'; 'finding out where children are,' and 'informing when to expect me home') for using your mobile phone to facilitate family/household coordination?" The greatest importance is attached to information about the timing of the arrival at home (81 percent) and arranging to meet with other family members (82 percent), while planning meals is rated as either "very important" or "important" by just a third of the respondents. Among parents, "arranging to deliver goods or children" and "finding out where children are" is rated as important by 63 percent and 58 percent, respectively.

While the use of the mobile for coordination is common, does this feature of the mobile phone necessarily override the separation between work and home? Under what circumstances will users attempt to control the flow of workplace demands through the walls that divide home and work? A key feature of work/life balance is the practice of taking holidays, away from both the workplace and the drudgery of home. The mobile phone is uniquely designed to function independently of location, so the notion of being "out of touch" while away on holiday no longer applies automatically. Mobile-phone users can now choose whether to stay connected or leave the phone behind. Interestingly, we found that the population of workers in our study is evenly divided between those who do take their phone on holiday to talk to work colleagues and those who don't. However, when this result is broken down by gender, it is apparent that men (51 percent) are almost twice as likely as women (31 percent) to be using their mobile phone to talk to their work colleagues while on holiday. It appears that women are more concerned than men to prevent the encroachment of their work into holiday time. Managers are the most likely (59 percent) to take their phone on holiday to conduct business, whereas only 30 percent of clerical workers do the same.

It is clear from these findings that people do understand the mobile phone as a technique for control over the interaction between work and personal life. Sometimes they use it to erect a boundary between home and work, while at other times they use it to intensify connection either with work or with family. Indeed, one of the principal uses of the mobile phone is to strengthen ties with kin and close friends at a distance. Respondents were asked "How important are the following in maintaining contact with your extended family?" and invited to rate various communication modalities on a five-point scale, ranging from "very important" to "very unimportant."

The mode of communication respondents consider most salient for maintaining contact with extended family were, in order of importance, the landline (83 percent), face-to-face visits (76 percent), the mobile phone (66 percent), followed by e-mails (61 percent), texting (48 percent), and then a large gap to the traditional modality of letter writing (23 percent) and the newest technologies of voice-over-Internet protocol (VOIP) (16 percent). Although the mobile phone is a much more recent innovation than the landline, it has already become a crucial tool for maintaining intimate ties, since nearly two-thirds of our respondents rate this function of mobiles as either "important" or "very important."

Indeed, communication in itself has become more central to intimate relationships with the new emphasis on reflexive self-identity, characteristic of modernity.[33] Rather than a specific informational content being the crucial element of making a mobile phone call, in many cases the call itself is constitutive of the relationship. Regardless of the communication modality, women are more likely than men to consider maintaining contact with family "very important." A good example is the use of the landline for keeping in touch with relatives. In our study we found that 86 percent of women consider that the landline is either "important" or "very important." Interestingly, nearly two-thirds of the women who regard the landline as a useful way of maintaining contact said it was "very important." The same pattern holds for mobile phones and e-mails. This is consistent with the literature on the gendering of the telephone that has demonstrated that maintaining kinship relations is traditionally a task undertaken by women.

Keeping in touch while physically apart is a marker of intimacy. If constant connection is the main quality afforded by mobile modalities, then might not this property be also allowing intimacy at a distance? Through the mobile phone, people can be apart and yet very close. In order to gain some insight into this possible use, we asked respondents: "If you and your partner are routinely apart for more than a day at a time, how important is the mobile phone in maintaining the quality of your relationship?" Approximately three-quarters of both men and women consider the mobile phone to be either very important or important in maintaining the quality of their relationship while geographically separated.

These findings demonstrate how the mobile phone's capacity to offer perpetual contact is associated with newer and more intense forms of connectedness. They tend to confirm Nicola Green's view that time spent using communicational devices makes relationships durable and continuing, rather than fragmented, which has been the conventional wisdom of the literature on ICTs.[34] The possibility that mobile technologies afford novel opportunities for deepening strong ties and making place irrelevant stands in direct contrast to the view that perpetual contact encourages work problems to colonize the social spaces and times once reserved for family life. Like other digital technologies, mobile phones have complex and contradictory effects.

CONCLUSION

Standard sociological readings of ICTs tend to give primacy to social relations, treating them as existing prior to and outside the intervention of technology. The STS approach adopted here, however, treats the technical as part of the constitution of the "social." ICTs are not only instruments of communication, but also create and reshape relationships between people, altering the basis of social interaction. The performance of family and intimacy is achieved in relation to the domestication of sociomaterial artifacts. The connectedness fostered by mobile modalities may be creating novel forms of intimacy, altering the very character of personal life, and changing the dynamics of work. In other words, as with other technological innovations, mobile devices may be ushering in a range of new social practices, communication patterns, and corresponding ways of life. It may thus be time to reformulate our analysis of work/home permeability in terms of how family practices are being mutually configured by mobile technologies.

We have seen that people use the mobile phone to provide a continuous pattern of interpersonal communication. These phone practices do not simply substitute or compensate for other forms of mediated and face-to-face interaction. Rather, they provide a way to combine connected relationships, redrawing the divisions between absence and presence. By implication, this also involves disrupting and rearticulating the spheres of public and private. The mobile phone promotes connectedness independent of space, providing a private realm for affective relationships among family members and friends. It significantly increases people's capacity to maintain intimacy at a distance and over the course of a normal day.

Much has been written about how mobile communications are blurring the boundaries between home and work life. We have argued that this discussion has been handicapped by the idea that the division between home and work is binary and fixed, and experienced at separate and distinct locations and times. By examining the ways in which mobile communications transgress the spatial and temporal basis of these traditional boundaries, we can begin to reconceptualize the complex issues associated with the shifting work/family interface. The sociological notion of family practices highlights a view of the family as actively constructed through the day-to-day activities of its members. These practices do not necessarily take place in the time and space conventionally designated to do with family, that is, the home. Indeed, family life is lived and maintained through the communicative practices of its members.

The ubiquitous use of the mobile phone for "staying in touch" with intimates, regardless of location and formal working hours, highlights this in new ways. The mobile illustrates the dangers of assuming that growing individualism and choice, said to be so characteristic of late modernity, results in social atomization and the ascendancy of a particular type of fragile intimate relationship.[35] Indeed, aspiring to warm, caring relationships may be an increasingly common feature of the intimate landscape, precisely because these relationships serve as an apparently attainable source of meaning in an anomic and fragmented world. While a mobile phone is a highly individualized technical artifact that people typically personalize to reflect their identity and sense of self, its capacity to offer perpetual contact may be associated with newer and more intense forms of connectedness. Rather than fragmenting time, our study suggests that mobile-phone practices are strengthening and deepening personal relationships, building durable social bonds.

NOTES

1. Cynthia Fuchs Epstein and Arne Kallenberg, eds., *Fighting for Time: Shifting Boundaries of Work and Social Life* (New York: Russell Sage Foundation, 2004); Arlie Russell Hochschild, *The Time Bind: When Work Becomes Home and Home Becomes Work* (New York: Metropolitan Press, 1997).
2. Sue Falter Mennino, Beth A. Rubin, and April Brayfield, "Home-to-Job and Job-to-Home Spillover: The Impact of Company Policies and Workplace Culture," *Sociological Quarterly*, 46 (2005): 107–35.
3. Janet Smithson and Elizabeth H. Stokoe, "Discourses of Work/Life Balance: Negotiating 'Genderblind' Terms in Organizations," *Gender, Work and Organization*, 12 (2005): 147–68.
4. Alan Felstead, Nick Jewson, and Sally Walters, *Changing Places of Work* (Hampshire: Palgrave Macmillan, 2005); Carol Kaufman-Scarborough, "Time Use and the Impact of Technology. Examining Workspaces in the Home," *Time & Society*, 15 (2006): 57–80.
5. Barry Brown, Richard Harper, and Nicola Green, eds., *Wireless World: Social, Cultural and Interactional Issues in Mobile Communications and Computing* (London: Springer Verlag, 2002); James E. Katz and Mark Aakhus, eds., *Perpetual Contact: Mobile Communication, Private Talk, Public Performance* (Cambridge: Cambridge University Press, 2002); Rich Ling, *The Mobile Connection: The Cell Phone's Impact on Society* (Amsterdam: Elsevier, 2004).
6. Noelle Chesley, "Blurring Boundaries? Linking Technology Use, Spillover, Individual Distress, and Family Satisfaction," *Journal of Marriage and Family*, 67 (2005), 1237–48; Linda Duxbury, Ian Towers, Christopher Higgins, and John Ajit Thomas, "From 9 to 5 to 24 and 7: How Technology Redefined the Work Day," in *Information Resources Management: Global Challenges*, ed. Wai Law (Hershey, PA: Idea Group Publishing, 2006); Nicola Green,

"On the Move: Technology, Mobility, and the Mediation of Social Time and Space," *The Information Society,* 18 (2002), 281–92.

7. Ling, *Mobile Connection.*
8. Nancy Folbre and Michael Bittman, *Family Time: The Social Organisation of Care,* (London: Routledge, 2004); Barbara Pocock, *The Work/Life Collision* (Sydney: The Federation Press, 2003); Barbara Pocock, *The Labour Market Ate My Babies: Work, Children and a Sustainable Future* (Sydney: The Federation Press, 2006).
9. Alison Morehead, Mairi Steele, Michael Alexander, Kerry Stephen, and Linton Duffin, *Changes at Work: The 1995 Australian Workplace Industrial Relations Survey* (South Melbourne: Longman, 1997), 118.
10. Australian Communications and Media Authority (ACMA), *Telecommunications: Performance Report 2004–05* (Melbourne: ACMA, 2005).
11. BuddeComm, "Australian Telecommunication Market Reports." In http://www.budde.com.au/reports/Category/Australia-7.html, accessed 22 May 2007.
12. Christena E. Nippert-Eng, *Home and Work: Negotiating Boundaries through Everyday Life* (Chicago: University of Chicago Press, 1995).
13. David Morley, *Home Territories: Media, Mobility and Identity* (London: Routledge, 2000), 87.
14. Ling, *Mobile Connection,* 59.
15. Duxbury et al., "From 9 to 5 to 24 and 7."
16. Chesley, "Blurring Boundaries?"
17. Joseph G. Grzywacz, David M. Almeida, and Daniel A. McDonald, "Work-Family Spillover and Daily Reports of Work and Family Stress in the Adult Labor Force," *Family Relations,* 51 (2002): 28–36; Sheldon Zedeck, *Work, Family and Organizations* (San Francisco: Jossey-Bass, 1992), 20.
18. David Morgan, "Risk and Family Practices: Accounting for Change and Fluidity in Family Life," in *The New Family?,* ed. Elizabeth B. Silva and Carol Smart (London: Sage, 1999).
19. Mary Douglas, "The Idea of a Home: A Kind of Space," *Social Research,* 58 (1991): 287–307.
20. Douglas, "The Idea of a Home," 306.
21. Ling, *Mobile Connection.*
22. Ed Hackett, Olga Amsterdamska, Mike Lynch, and Judy Wajcman, eds., *The Handbook of Science and Technology Studies* (Cambridge, MA: MIT Press, 2008).
23. Bruno Latour, "Where Are the Missing Masses? The Sociology of a Few Mundane Artifacts," in *Shaping Technology/Building Society: Studies in Sociotechnical Change,* ed. Wiebe Bijker and John Law (Cambridge, MA: MIT Press, 1992); John Law and John Hassard, eds., *Actor Network Theory and After* (Oxford: Blackwell, 1999).
24. Leslie Haddon, *Information and Communication Technologies in Everyday Life* (Oxford: Berg, 2004).
25. Christian Licoppe, "Connected Presence: The Emergence of a New Repertoire for Managing Social Relationships in a Changing Communication Technospace," *Environment and Planning D: Society and Space,* 22 (2004): 135–56.
26. Chantal De Gournay, "Pretense of Intimacy in France," in *Perpetual Contact: Mobile Communication, Private Talk, Public Performance,* ed. James E. Katz and Mark Aakhus (Cambridge: Cambridge University Press, 2002), 201.
27. Ann Moyal, "The Gendered Use of the Telephone: An Australian Case Study," *Media, Culture and Society,* 14 (1992): 51–72.

28. Lana Rakow, *Gender on the Line: Women, the Telephone, and Community Life* (Chicago: University of Illinois Press, 1992).
29. Lynn Jamieson, *Intimacy: Personal Relationships in Modern Societies* (Cambridge: Polity Press, 1998).
30. Judy Wajcman, "TechnoCapitalism Meets TechnoFeminism: Women and Technology in a Wireless World," *Labour & Industry*, 16 (2006): 7–20.
31. ARC Linkage Project (LP0667674), *The Impact of the Mobile Phone on Work/Life Balance*, Judy Wajcman, Michael Bittman and Paul Jones; Research Fellows, Jude Brown and Lynne Johnstone.
32. Ling, *Mobile Connection*.
33. Anthony Giddens, *The Transformation of Intimacy: Sexuality, Love and Eroticism in Modern Societies* (Cambridge: Polity, 1992).
34. Green, "On the Move."
35. Jamieson, *Intimacy*; Carol Smart, *Personal Life: New Directions in Sociological Thinking* (Cambridge: Polity Press, 2007).

3 Gender and the Mobile Phone

Leopoldina Fortunati

INTRODUCTION

As the mobile phone grows into a multimedia device and becomes mobile media, one dominant issue has continued to haunt the device throughout its various permutations and incarnations. The rise of mobile media could be clearly read in terms of arising and existing gender inequalities that stretch across many notions of mobility and immobility. The purpose of this paper is to use the lens of the mobile media to reflect upon constructions of gender over the last couple of decades. One of the central arguments of this paper is the need to problematize categories such as gender, generation, domestic sphere, social relationships, family, and so on. I argue that often these notions are used in a statistical, aseptic way. With the plenitude of sociological debates around notions of mobility spanning at least a decade, I argue that we have an expansive and sophisticated material in which to expand upon debates around gender. In particular, I will focus upon the intersections between questions of gender, generations, and especially domestic-sphere categories. I argue that it is time to develop research that is specifically designed to study the role, meaning, representations, models, and practices of use of the mobile phone beginning from women's life conditions.

Even a cursory survey of the empirical work on gender and the mobile phone to date generally shows—with few exceptions—research methodologies that unproblematically take the masculine situation as a central precept. When gendered performativity is analyzed, it is often done so in comparison—the "feminine" is defined in terms of the "masculine." Or, as it happens more often than not, there are seemingly no significant differences between the behavior of men and women. This lack of differences, however, is rarely discussed: thus leaving a series of questions unanswered. What does it mean to say that there are not significant differences between men and women? Does it mean that the differences between men and women are diminishing or does it mean that differences exist but they become indiscernible in current research agendas? To rectify this dilemma, the first step might be to alternate our direction of inquiry. Instead, we should ask questions such as: What types of experiences and agencies does the mobile phone give to women? Does the mobile phone serve, for instance, to reduce time, fatigue, and costs related to the double work (housework and the waged work) or not? Does the mobile phone serve to improve on the whole the social condition of women or not?

In regards to the methodological issues, there is an urgent need for a reconceptualization of the research tools, by which we mean not only specific techniques, but the whole level of methodological resources. For example, has the explicatory and predictive capacity of the gender variable changed? It is impossible to answer this question right now: what we can do now is use these research questions to establish a future research agenda. Concepts, indicators, variables, techniques, and tools should be reconceptualized, reviewed, and replanned, in order to be able to deal effectively with this particularly elusive object that is social change connected with gender.

GENDER, GENERATION, AND THE MOBILE PHONE

In the first wave of research on the mobile phone, both women and men were explored in a coy manner in regards to their sociological complexity as social actors. This initial wave of research was limited in that it rarely problematized the gender category from a sociological point of view. Women, like men, have been considered, in general, as an operational category simply characterized by the self-definition on the part of respondents. But gender category is, of course, more complex than this: it is a social construction and representation; it is a model or, better, a series of models; it is a set of individual, familiar, and social roles; it is a body of images and figures.[1] Furthermore, implicated in the issue of gender are also the underlining modes of gendered labor and politics of time/space, particularly prevalent in discussions around reproductive labor.[2]

A second limitation with this initial research was that although the majority of these research projects looked at the diffusion, adoption, and use of the mobile phone in the domestic sphere and everyday life—as exemplified by the domestication approach[3]—these two notions of gender and domesticity have been rarely problematized. They too remained as background notions, like narrative tools used to give an idea of a specific framework in which the use and appropriation of the mobile phone was taking place, but which were never defined on a sociological level, that is, in their economic and social functionality in respect to the whole economic system. A third shortcoming of this first decade of research was perhaps the lack of historicization of social phenomena connected to the use of the mobile phone. The diffusion and appropriation of the mobile phone has taken place vis-à-vis specific generations of adolescents, children, and women roughly from 1995 till now. And it is necessary to know much more about them from a social, economic, and political point of view.

Thus, to make the next wave of research more sophisticated, first of all we need to expand upon the question of gender in the light of the recent debates on this topic. When we observe and try to understand differences in social behavior, we should take in account that the attribution to a gender needs to be understood as a socially constructed notion that is different from a biological sex. Poststructural theorists such as Judith Butler[4] and Kate Bornstein[5] argued that we are constantly in the process of *becoming* a gender rather than *being* a gender. So the social "rules" shape gender change on the basis of the continued transformations that the social construction of "masculinity" and "femininity" undergoes. According to Butler, gender is not a primary or "real" category but an attribute, a set of secondary narrative effects that are both performed and regulated.

So in our research we should bear in mind that the category of "gender"—like identity—is highly dynamic and subject to sociocultural forces. In particular, this reconceptualization of gender is enlightening, especially in the context of the relatively underexplored "feminine" approach to technology[6] in the studies of the mobile phone.

This understanding of gender allows a closer examination of stereotypes around gendered usage of technology. Moreover, through this understanding of gender we can gain greater acuity into the rise of the mobile phone ties to overtly gendered formations. There are various levels in the role of gender, and notions of the feminine operate within mobile media culture. Firstly, the role of the "feminine" is integral in the design process. The first mobile phones were, like many other information and communication technologies (ICTs), aimed at young adult white and affluent men. From the start, mobile phones were conceived as a tool for male yuppies.[7] Secondly, we need to factor in time issues: when a technology is conceived for a particular segment of the population, the other potential buyers need time to understand how they can adapt the technology to their needs. It is these users (second and third wave of users) who generally redesign and reconfigure the device: a phenomenon marked by gender and modes of feminizing the technology.

Lastly, it depends on the demographic of women: age is a key variable to predict to what extent feminine behavior differs significantly or not from that of men. Young generations of women now show very modest parameters of difference. Often, as in the case of mobile phone practices, these differences are expressed in the form of disadvantage in comparison to their male counterpart.[8] Young women, adolescents, and children were able to reduce in a very short time the "delay" that has accompanied in general the adoption of mobile phones (and also computer and internet) by women.[9] In a few years young women had surpassed the usage by young men but also overcame them. This social dynamics should be studied with great attention—taking into account that that girls' and female adolescents' power is weakened by the fact that adult and mature women have less social power, which, in turn, influences the deployment of certain technologies. The minor social power of women in general inevitably tends to lower expectation for younger women, and this element might explain why age, in many situations, is a factor of social aggregation that overrides the significance of gender, class, ethnicity, and so on.

On the whole, women's approach to the mobile phone developed in a way not so different from that expressed towards the old communication technologies and mass media. The schema was always the same: men opened the way and then women slowly followed, trying to adapt to their interests and needs the new technology present in the market. However, this was not the case with domestic appliances which were expressly designed for women; and with which women are at ease, while men are, in many cases, incompetent and hostile.

Gender has already been widely recognized as a fundamental dimension in the whole negotiation process which takes place between society and the technological world, made up by several actors—such as innovators, policymakers, users, operators, and manufacturers, to name but a few. As Ellen van Oost has argued, "the process of shaping gender and technology are closely intertwined. One can even speak of a 'co-construction of gender and technology.'"[10] This means that it is very important to develop a robust analysis of the role and position women, in order to register and understand how men and women coconstruct the different technologies in a specific ways.

Women, for instance, tend to focus upon the physicality and functionality of the mobile phone. Women have negotiated the design of the keyboard and in general of the entire body of the mobile phone with operators and manufacturers. The mobile phone has become cute, colored, and fashionable. The introduction of "feminine" services and applications has increased the possible modes of gender performativity from calendars for menstrual cycles, calories calculations, camera phone genres,[11] and so on. For this reason it is particularly pertinent to understand whether the new wave of mobile phone (Universal Mobile Telecommunications System [UMTs] and 3G technologies)—which are more technologically and aesthetically sophisticated than the current ones—will attract women faster compared to Total Access Communication System (TACS) and Global System for Mobile, originally Groups Spécial Mobile (GSM). This last issue is specifically important, because we need to figure out if, and to what extent, the feminine approach to innovations—and empowerment—is changing.

This reconceptualization of categories is necessary not only in respect to the notion of gender but also *generation*. Although a large part of the studies on mobile communication focused on adolescents and teens, that is, generations born in the second part of the seventies and in the eighties, the notion of generation has rarely been challenged. Research on "age" implicitly related to "youth"; anything not encompassed by youth was often seen as boring and not worthy of research. What type of generation do we talk about? In much of the research, the category of generation is implicitly treated as an age cohort.[12] In the majority of these studies the notion of generation was used instrumentally without any specific reflection. The minimalist concept of generation that emerges here is added proof that a very fundamental question has been avoided: what are the distinctive social, cultural, and political identity of this generation made up by teens and adolescents in the nineties?

Unquestionably, a reconceptualization of the notion of generation would provide new insight into why, historically, this was the first generation to be impacted by mobile technology. From the outset, the mobile phone has been able to attract adolescents much more rapidly than adults. Adolescents and teens of this generation were in the majority of cases only children. Not having any brother or sister, they probably saw in the mobile phone a powerful tool to guarantee to themselves in the family the virtual presence of somebody of the same age. Through the mobile phone, the communicative fluxes with their peers could continue at home. It was looking at this phenomenon that we talked of the social construction of virtual brotherhood and sisterhood.[13] Given that copresence in the peer community was possible for this generation of adolescents and teens only in a fragmented way, it is easy to understand why mobile communication became—for these generations—the bridge that allowed for the continuity of their peer relationships in the different places of everyday life (school, family, sport, and so on).

But this, of course, is only an aspect of the problem. Why, for instance, was this generation able to win more power inside the family than in the political and social word? It is a matter of fact that while these adolescents were able to become the main explorers and users of mobile technology, they didn't take an analogous initiative in changing social and political rules, attitudes, and behavior towards them.[14] We need to understand better their situation, the specific factors involved. For example, consider factors such as increasingly inclusive schools and a hostile job market that offered little to these new generations. This situation, in part, explains why this generation produced moments of "Durkheimian effervescence," like the no-global movement.

Furthermore, the generation of the nineties in Eastern countries and in Germany lived through the fall of the Berlin wall and the redesign of the political Europe, experimenting with a new sense of transnational identity. In other words, this mobile generation is characterized not only by the experience and practices of the mobile phone use but also by broad social, economic, and political changes in the family, in the relationships among peers and between parents and children, in the labor market, in the educational world, and so on. Here are but some examples of the many reasons why the uptake of mobile technologies was so rampant amongst this generation, thus establishing a template for future generations. It would be important that next wave of research could give back to these generations a high sociological thickness that helps us to provide answers to the several open-ended questions left hanging.

DOMESTIC SPHERE AND MOBILE PHONE

Another notion that is necessary to challenge is that of the domestic sphere, the setting where the greatest numbers of researchers have chosen to analyze mobile-phone use. We need to understand this issue better in order to avoid the domestic sphere remaining as a minimalist and instrumental context. How can we define the sphere of everyday life from a sociological point of view? Of course, feminist tradition can offer much acuity upon the assumptions about gendered space and labor inequalities. The classics of feminist thought have analyzed and depicted everyday life as the place made up by family, house, housework, marriage, birth, life, and death and where the most precious commodity for our societies is built: the labor force, the human being.

The plurality of feminist approaches, whilst being pivotal, has had many problems in the distribution within the academic world. Here centuries of hostile tradition towards everyday life have created strong inertia and resistance to this discourse. Romantic literature (let alone classical) and a large part of the philosophical and economical tradition condemned almost without appeal everyday life to represent the place of the ugly, trivial, insignificant, unproductive, the mere survival. The aporias of the culture of the eighteenth and nineteenth centuries have produced an opposition between "what happens every day" and extraordinary or, at least, significant events that happen in the public sphere.[15]

Henri Lefévre[16] and Michel de Certeau[17] analyzed with great inspiration everyday life; more recently, the late Roger Silverstone[18] looked at this sphere trying to understand the logic that governs it, proposing the notion of "moral economy." Remembering his unflagging work, we recall that he created inspiration to study the role of ICTs in the domestic sphere,[19] proposing the "domestication theory" in order to understand the integration of ICTs in everyday life.

Here I would like to merge together specific feminist analysis of the domestic sphere and debates on social capital. These two branches of debate might help us to reconceptualize the notion of the domestic sphere in a more complex way. As I have argued elsewhere, the reproductive sphere is made up by family, housework and care work, leisure, social interactions, communication, organization, education, entertainment, affects, love, and sex.[20] Analyzing the reproduction sphere by moving beyond Marxian categories, in earlier research I have showed how the sphere of everyday life, although it represents itself as a sphere that naturalizes process, is as much productive for the general socioeconomic system as the sphere where commodities and services are produced and contributes in fundamental terms to the valorization process, to the production of surplus value.[21] The debate on social capital allows us to enlarge this notion of reproductive sphere and include also information, knowledge, civil engagement, and culture. Today it is not unusual to find scholars discussing human, cultural, or social capital within the context of the mobile phone.[22] However, the variety of definitions of this term and the variety of disciplines implicated in the debate (sociology, education, economics, business, political science, and organizational behavior) mean that we need to clarify, even if briefly, a working definition.

For Marx, workers were the only elements of the capitalist process that were able to produce a variation (in term of augmentation) of the value, while machines and instruments of labor could only release their value. For this reason, Marx called workers "variable" capital and machines "fixed" capital. Marxian analysis was unfortunately restricted only to the sphere of commodities production: the factory world. In 1953, Selma James gave a vivid description of how this "variable capital"—that is, people—was reproduced essentially by women in the domestic sphere.[23] She showed how—and at what personal price—women contribute to gross domestic product (GDP). Her text, written in the height of the cold war, is particularly important because it describes with lucidity the contradictions and the real feelings of women who were divided between the house and the place of work in a town, Los Angeles, at that time with a strong internal immigration.

In 1961, Jane Jacobs wrote about the value of informal interpersonal relationships.[24] Although she never used the term *social capital*, Jacobs underlined how these relationships are crucial for the functioning of complex and highly organized societies. Jacobs understood that value production, being a process, could not stop inside the walls of the factory and that outside the factory, in the reproduction world, an analogous process had to exist. She saw this value, this "capital," as the amount of informal social relationships. The limit of her vision was that she stopped at this stage, without looking at the work process which was behind interpersonal relationships and which has in turn its specific "workers."

Three years later, Gary Becker argued that education, training, and knowledge of individuals in general are crucial factors for the productivity of nations and introduced the notion of human capital.[25] With the expression *human capital*, he stressed the importance of these elements for the valorization process. He argued that in industrialized countries human capital would represent 80 percent of the whole wealth of these countries. In Becker's approach, human capital is produced by family efforts and public investments in the education system. However, by focusing generically on family efforts, this prevented him from recognizing the specific role played by women on this. "Human capital" is, in a sense, produced by means of an unwaged labor—that is, housework and care work. In the domestic sphere, housework produces not only education and knowledge but also, more largely, the whole subsistence of new generations—starting from their existence, birth, and bodies.

To complete Becker's analysis on valorization, one should include also all the labor involved for the entire existence and subsistence of people, not only the segment of education. In the seventies Maria Rosa Dalla Costa[26] took up where Selma James's analysis finished. She put together all the pieces of the puzzle—housework, pregnancy, childbirth, abortion, double work, love, sex, education, social relationships, and so on—to outlined a coherent framework in which all these tessera came together. This site, the domestic sphere, was analyzed as the sphere where the most precious commodity, the labor force in the form of the individual, is produced and reproduced everyday.

Influenced by feminism, Michel Foucault[27] introduced another point of view on the issue of valorization, in the domestic sphere, by focusing on population. He underlined the value represented by population for a country and analyzed the political mechanisms allowing the control of population fluxes, dimensions, and changes through the notion of biopower. According to Foucault, in the eighteenth century, biopower developed in two directions: on the one hand, it introduced practices of discipline and education among individuals and, on the other hand, it tried to control population, as a machine to produce wealth, goods, or other individuals. Especially, in the second half of seventeenth century, continues Foucault, a series of issues, which were connected to this new value recognized to population, began to be debated: habitat, material and social conditions of urban life, public hygiene, changes in the relation between birthrate and death rate. According to Foucault, it was clear that the capitalist state needed to regulate population fluxes in order to strengthen its power; sex and reproduction become fundamental elements in the political transformation of society. However, again, within Foucault's analysis the problem of production and reproduction of labor force was considered starting from the end, not from the beginning. The value of population was seen in fact as a given data, a "natural" output.

Pierre Bourdieu's [28] notion of social and cultural capital adds some new elements to this picture of everyday life. In particular, he saw social capital as "the aggregate of the actual or potential resources which are linked to possession of a durable network of more or less institutionalised relationships of mutual acquaintance and recognition."[29] This notion of social capital, which will be taken up later on by James Coleman[30] and Robert Putnam,[31] was a new opening towards considering social relations as something closely connected to capital and its valorization. However, this definition, I argue, is incorrect, because it refers to a kind of capital that belongs to individuals, whereas social relations and contacts represent instead one of the territories of the immaterial production that works as a means of valorization for the capital in the reproductive sphere.

Coleman broadened the concept of social capital, defining it as anything that supports and facilitates both individual and collective action. In his vision, networks of relationships, reciprocity, trust, and social norms generate social capital. Putnam enlarged further the concept of social capital, including civic engagement and talking about bonding and bridging social capital. Furthermore, he saw social capital as a resource possessed not only by individuals but also by collectives and communities. The debate on social capital has in time become very complex: Roberto Cartocci and Valerio Vanelli showed that the use of this concept is a kind of torsion of the Marxian notion of variable capital.[32] In any case, for our purposes it is sufficient to refer to the main important ideas developed inside this debate.

So how do all these themes relate to the domestic sphere? How do the ICTs enter in this working process? They enter as devices that are activated to support all the various activities that can be included under the wide umbrella of reproduction labor. This labor is made up by, in part, still a large amount of *material labor* (cook, shop, clean, wash, iron, etc.) but also by an increasing part of *immaterial labor*: organization, communication, information, entertainment, civil and political engagement, education, interaction, affects, love, sex. Material and immaterial labor converge to create value, a surplus value that is embodied in human beings. The other side of the variable capital is exactly this: the production of domestic capital.

This notion of domestic or reproductive capital includes, systematizes, and enlarges the notions of human and social capital. In this framework, the mobile phone, like the other information and communication technologies, works on the side of *immaterial reproduction*. Each of these technologies is adapted to support one or more aspects of the immaterial labor in the everyday sphere. Drawing from the current research, the mobile phone supports various forms of communication, information, organization, social relationships, contacts, civil and political engagement. It helps to carry out all these types of immaterial reproductive labor. The diffusion of the mobile phone can be seen as an increase in what Karl Marx calls "fixed capital"—that is, machines. Machinery, first accumulated at the specific sites of factories, is now spread and appropriated also in the domestic sphere. So the typical user of fixed capital is no longer the classic worker in a factory or firm but, rather, women and in general the ordinary citizen in his or her domestic sphere. In this sphere, the mobile phone might be considered as a *work tool for reproduction*. That is, a tool that supports and facilitates almost all the aspects of immaterial reproductive labor, which are increasingly complex and exponential in influence.

In other words, the mobile phone, as emblem of an extraordinary development of mechanization of immaterial labor, helps to rationalize the management of the immaterial reproductive sphere and to increase the production of value in this sphere. This device—and in general ICTs—have, on the one hand, strengthened and, on the other hand, transformed reproductive labor into an increasingly mediated, self-reproductive, and self-disciplinary form. Not only has it grown greatly in the domestic sphere, but also it has done this to the detriment of the material labor of reproduction, which, in part, has been suppressed or expelled from houses and subjected to outsourcing.[33] Furthermore, the logic that governs the sphere of reproduction of the labor force has been exported into the world of production of goods, where there is today a strong presence of immaterial, unpaid, scarcely defined legally, precarious work, and so on.

So as reproduction is extending its dominion over society today, the mobile phone has become also a strategic *tool of social labor*. But why do I speak of only women and not of the ordinary citizen? This is because immaterial reproductive labor occurs on different levels to all individuals, women and men as well as children and adults. In this sense, the carrying out of immaterial reproductive labor pertains to all individuals and citizens. Each individual has a double role in the domestic sphere, where he or she is producer and consumer at the same time. The quantity and the quality of reproductive labor are negotiated in different ways—according to the historical moment, social and political relationships—between genders and generations and different cultures. However, even recent researches show that there are those who consume more (children, adult males, the elderly) and those who continue to work more, even if less than once (women).[34] In sum, if we want to understand the socioeconomic role of the mobile phone, we need to revisit feminist debates around social capital. Only in this framework does the mobile phone shows its identity as a machine, which is set up and used as a work tool for immaterial reproductive labor within the social processes governing everyday life.

CONCLUSION

In this chapter I called upon the urgent need for mobile communication researchers to problematize the sociological categories that we activate in our empirical research. In particular, I have showed one of the possible ways to reconceptualize this terrain through the intersections between gender, generation, and domestic sphere. In regards this latter category, the domestic sphere, I have problematized this social space and "moral" economy in the light of a specific feminist tradition and debates related to social capital. The fertilization of these two branches of analysis has stimulated a new reflection on the socioeconomic role and meaning of the mobile phone and mobile communication.

I have argued that it is necessary to move to a new stage of research and analysis, stimulating the shift from a descriptive stage of the studies on mobile telephony to a second stage, a more complex paradigm which negotiates the new forms of labor and capital of postmodernity.[35] In this paper I have highlighted some of the theoretical and methodological limitations that have characterized the thirteen odd years of preliminarily mobile-phone research. In the diffusion of broadband technology and the convergence process that are occurring between Internet, pictures/video/television, and mobile phones, I believe we are still unprepared to understand how women and the different generations will react to mobile media. What remains apparent is that mobile media will undoubtedly intensify the gender and generational logics already at play. Thus, the challenge remains how women—of different generations—will build important moments of self-determination and self-valorization through the realm of mobile media.

In conclusion, this paper is homage to the efforts of all the women who, during the last decades, have been engaged in understanding and studying women's place and sense of being in the world. Although often lacking in resources, funding, academic positions, and so on, many women, such as all the female scholars I have cited in this chapter, have given a great contribution to understanding the meaning and the functioning of the domestic sphere, as well as the relation between women and technology, and showed that these themes are so crucial not only for women but for all.

NOTES

1. Judith Butler, *Gender Trouble: Feminism and the Subversion of Identity* (London: Routledge, 1990).
2. Judith Wajcman, *Feminism Confronts Technology* (Oxford: Polity Press, 1991).
3. Roger Silverstone, Eric Hirsch, and David Morley, eds., *Consuming Technologies: Media and Information In Domestic Spaces* (London: Routledge, 1992).
4. Judith Butler, *Gender Trouble*.
5. Kate Bornstein, *Gender Outlaw: On Men, Women and the Rest of Us* (New York: Routledge, 1994).
6. Merete Lie, ed., *He, She and IT Revisited: New Perspectives on Gender in the Information Society* (Oslo: Gyldendal, 2003); Cynthia Cockburn and Susan Ormrod, *Gender and Technology in the Making* (London: Sage, 1993).
7. Jon Agar, *Constant Touch: A Global History of the Mobile Phone* (Cambridge: Icon Books, 2003).
8. Joachim R. Höflich and Julian Gebhardt, eds., *Mobile Kommunikation. Perspektiven und Forschungsfelder* (Frankfurt am Main: Peter Lang, 2005).
9. Leopoldina Fortunati and Raimondo Strassoldo, "Practices of Use of ICTs, Political Attitudes among Youth, and the Italian Media System," in *New Technologies in Global Societies*, ed. Patrick Law, Leopoldina Fortunati, and Shanhua Yang (Singapore: World Scientific Publisher, 2006), 25–158.
10. Ellen van Oost, "Introduction," in *Designing Inclusion: The development of ICT Products to Include Women in the Information Society*, ed. Els Rommes, Irma van Slooten, Ellen van Oost, and Nelly Oudshoorn (Twente, Netherlands: University of Twente, 2004), 7.
11. Larissa Hjorth, "Snapshots of Almost Contact: Gendered Camera Phone Practices in a Case Study of Seoul, Korea," paper presented at *Cultural Space and the Public Sphere in Asia*, hosted by Asia's Futures Initiative, March 15–16, Seoul, 2006.
12. June Edmunds and Brian S. Turner, "Global Generations: Social Change in the Twentieth Century," *British Journal of Sociology*, 56(4) (2005): 559–77.
13. Leopoldina Fortunati and Annamaria Manganelli, "El Teléfono Móvil de los Jóvenes," in *Revista de Estudios de Juventud*, 57 (2002): 59–78; available (English version) at: www.mtas.es/injuve/biblio/revistas/Pdfs/numero57ingles.pdf.

14. Carlo Buzzi, Alessandro Cavalli, and Antonio de Lillo, *Giovani del Nuovo Secolo* (Bologna: Il Mulino, 2002).
15. Tzvetan Todorov, *L'Homme Dépaysé* (Paris: Éditions du Seuil, 1996), 125–36.
16. Henri Lefévre, *Critique de la vie Quotidienne*, vol. 3 (Paris: L'Arche, 1947–81).
17. Michel de Certeau, *L'Invention du Quotidien* (Paris : Éditions du Cerf, 1980).
18. Roger Silverstone, *Television and Everyday Life* (London: Routledge, 1994).
19. Roger Silverstone and Leslie Haddon, *Television, Cable and AB Households: A Report for Telewest* (Sussex, UK: University of Sussex Press, 1996).
20. Leopoldina Fortunati, "Immaterial Labor and Its Mechanization," *Ephemera: Theory and Politics in Organization*, 7(1) (February 2007): 139–57, http://www.ephemeraweb.org/journal/7–1/7–1fortunati.pdf.
21. Leopoldina Fortunati, *L'Arcano della Riproduzione* (Venezia: Marsilio,1981); English translation *The Arcane of Reproduction* (New York: Autonomedia, 1995).
22. Richard Ling, Birgitte Yttri, Ben Anderson, and Deborah DiDuca, "Mobile Communication and Social Capital in Europe," in *Mobile Democracy. Essays on Society, Self and Politics*, ed. Kristóf Nyíri (Vienna: Passagen Verlag, 2003), 359–74.
23. James Selma, "Woman's Place," in *The Power of Women and the Subversion of the Community*, ed. Maria Rosa Dalla Costa (Bristol, UK: Falling Wall, 1972).
24. Jane Jacobs, *The Death and Life of Great American Cities* (New York: Random House, 1961).
25. Gary S. Becker, *Human Capital*, 1st ed. (New York: Columbia University Press, 1964).
26. Maria Rosa Dalla Costa, ed., *The Power of Women and the Subversion of the Community* (Bristol, UK: Falling Wall, 1972).
27. Michel Foucault, *Histoire de la Sexualité: Tome—La Volonté de Savoir* (Paris: Gallimard, 1976).
28. Pierre Bourdieu, "Le Capital Social: Notes Provisoires," in *Actes de la Recherche en Sciences Sociales*, 3(31) (1980): 2–3; Pierre Bourdieu, "Forms of Capital," in *Handbook of Theory and Research for the Sociology of Education*, ed. J. C. Richards (New York: Greenwood Press, 1983).
29. Pierre Bourdieu, "Forms of Capital," 249.
30. James Coleman, "Social Capital in the Creation of Human Capital," *American Journal of Sociology*, 94 (1988): 95–120; James Coleman, *Foundations of Social Theory* (Cambridge, MA: Harvard University Press, 1990).
31. Robert Putnam, *Bowling Alone: The Collapse and Revival of American Community* (New York: Simon & Schuster, 2000); Robert Putnam, "The Prosperous Community: Social Capital and Public Life," *American Prospect*, 4(13) (1993): 35–42.
32. Roberto Cartocci and Valerio Vanelli, "Atlante del Capitale Sociale," *Sociologia del Lavoro*, 102 (2006): 169–91.
33. Arlie Russell Hochschild, *The Time Bind: When Work Becomes Home and Home Becomes Work* (New York: Metropolitan/Holt, 1997).
34. Jens Bonke, *The Modern Husband/Father and Wife/Mother—How Do They Spend Their Time?* www.oif.ac.at/sdf/sdf_05_2004.html.
35. For me, this new stage began with the workshop "Wireless Communication Workshop: Inspirations from Unusual Sources," organized by Christian Sandvig, Harmeet Sawhney, and Michael Traugott, University of Michigan, Ann Arbor, 26–29 August 2004.

Part II

Youth, Families, and the Politics of Generations

4 Children's Broadening Use of Mobile Phones

Leslie Haddon and Jane Vincent

While the first wave of youth and mobile-phone studies focused, understandably, on communications, the aim of this book is to reflect upon ongoing, media-related innovations related to this device. The cameraphone has become increasingly ubiquitous, which has already prompted a separate strand of research.[1] MP3 players on mobiles phones have become more common, as has the means to transfer files between mobiles phones and other devices via Bluetooth and infrared. These developments facilitate the role of the mobile phone as a digital wallet, as a digital album in relation to images, and as a digital music collection in relation to sounds. Mobile phones have increasingly been able to access the Internet and now television. Hence, at this juncture one might ask how the proliferation of media on the mobile phone has fitted into the individual and collective lives of children and youth.

This research[2] builds upon an earlier study.[3] In 2007, the researchers invited children to participate from the same schools and areas as before, all located in the south of England. Three focus groups comprising six to nine children were held in each of two secondary schools. In addition, forty-eight children,[4] including some of the focus group attendees, kept a twenty-four-hour diary record of their activities and use of the mobile phone. Lastly, six of the boys and six of the girls, who had not participated in a focus group but who had filled in the diaries, were then interviewed for about one hour about their diary and about their experience of mobile phones and other technologies. All the respondents were from the age groups eleven to twelve, thirteen to fourteen, and fifteen to sixteen.[5]

It is important to frame this research and analysis in various ways. First, one principle of the domestication approach,[6] discussed in the chapter by Judy Wajcman, is that British studies using this framework often attempted to be holistic in terms trying to understand technology in relation many aspects of people's lives: the financial circumstances, their hopes and aspirations, their leisure activities, their temporal organization of their lives, and so on. That is very time-consuming research and could not be achieved in this research, in what was a relatively small, short-term study. However, we can at least take a holistic view of young people's technological options rather than just looking at the mobile phone in isolation.[7] The young people in this study, originally texting, making voice calls (fixed line or mobile), and using e-mail, have in recent years increasingly embraced instant messaging (IM) on MSN as well social networking sites such as MySpace and Bebo,[8] where one can leave messages. In the case of newer media on mobile phones, the cameraphone is often competing with children's separate digital cameras. The MP3 function sits alongside many children's separate MP3 players, including iPods. Internet access on the mobile has to be assessed against children's access via, normally, the PC, while the TV on the mobile is another option compared to the family or personal TV that they can view. Lastly, we have games on mobile phones, competing with more sophisticated portable consoles. In sum, we need to be sensitive to the plethora of devices or communications options with similar or related functionality to that of the mobile phone that is making the world of young people full of more and more technological choices.

Given that much of the discussion will be about these choices, a second observation is that they are not meant to imply uncritical assumptions about youth and technology more generally. The first of these is that youth are at the cutting edge of innovation. One can say that what many of the young people in this study shared was more disposable time to experiment with technology than many adults, and the literature on youth often points to the greater influence of peer networks at this life stage, which may have a bearing on interest in exploring technology. However, even within this study there was a great diversity in terms of any such interest, the sophistication of their equipment, and hence the media options available to them as well as their own media competencies. And, as we shall see, these young people did specific things with their mobile phones that may or may not carry over into adult life. The other popular claim sometimes heard, which also unduly homogenizes youth, it that "this generation is different because. . . ." The interesting version of this claim at this moment in time concerns the cohort of youth who experienced mobile phones and texting when they first appeared. A few years ago one could ask how the next generation of youth, now growing up with mobile phones around, might have a different experience, one where they were not the innovators. However, even in this study we can see how the older teenagers, pioneers in the sense that they were the first to get mobiles several years early, had to deal with the new media on mobiles at the same time as the younger teenagers were doing so. The technological environment for both had shifted.

The third and final observation concerns the fact that this is a British study. The literature on the social construction of childhood[9] points to the fact that our expectations of what children should and can do, what they should be like, can change over time and can differ across cultures.

Indeed, technologies can become implicated in changing expectations of children, as was noted in discussions of what is the appropriate minimal age for children to have mobiles.[10] Hence, although there may be some commonalities in the experiences of young people as shown over the next chapters in this book, it has to be borne in mind that the different national contexts will mean that there will be some differences, sometimes nuanced, sometimes more outstanding, in the experiences of the young people being studied.

COMMUNICATIONS

One key consideration influencing the amount of calling and texting, which was equally true of the previous study, was the cost of communication.[11] Almost unanimously the young respondents in this study were cost conscious, reflecting the realities of their financial situation and dependence. The most common arrangement was for parents to set an allowance, usually credit on pay-as-you-go systems, but occasionally contract—and if they went over the limit the children would be expected to pay the difference. Even the "heavier" users tried to avoid going over the limit and some managed their finances better than others did. As Carol explained, when she sometimes ran out of credit, "I call people from the house phone, or I just keep on at my mum" to give her some more credit. For Carol, she wanted her mother to pay for her phone bills so that she could spend her own money on other things like clothes.

For those less able to regulate their communications, like Tim, the patterns of communication often followed a cyclical monthly binge—initial heavy use at the beginning of the month, resulting in all the credit being used up quickly, meant that the rest of the month the phone was underused. Although some informed peers of their situation, occasionally advising them to use IM instead, this pattern could have negative consequences. Mark's friends regularly gave up trying to contact him whenever he went through periods when he could not reply to texts. This cost consideration meant trade-offs: even when they might prefer the privacy and convenience of calls on their mobile phone, respondents sometimes used the landline. For example, although Shannon's friends tended to call her on her mobile, she thought it more prudent and "cost free" (from her viewpoint) to use the house phone to call friends.

For the most part communications were after the school day and at weekends. Voice calls made away from home were short communications with parents to do with picking up children, a summons to come back home, or letting parents know plans or where they were. And there were mobile calls to and from (mostly close) friends. As in other studies, texting was mainly to peers, which for some meant the routine of checking for texts in the morning. Some broke the rules and texted during the school breaks, either to keep up to date with friends in other schools or as something to do, a point we shall return to later. "It's more of the dare of actually doing it in school . . . than actually receiving it" Paula (14) (Focus Group 1).

Over time the pattern of mobile communications changed, but this is subject to mixed influences. A number of children felt that they communicated more as they grew older and increased their circle of friends and/ or there were more activities in which they were involved. However, the rise of alternative communication channels, especially IM, had had some impact on texting. Various participants described how instant messaging may have been around a while, but it had become more popular in their social circles during the last few years. Compared even to texting it had the merit of quickness (typing with the keyboard), being "more like a conversation," but chiefly it was free. Some, like Martin and Harriet, thought that their texting had declined because of the switch to instant messaging (IM). And Carol, on contract and one of the few to use the Internet on the mobile phone, used instant messages sent from the mobile's Internet function to interact with peers since the family computer was out of action. Not all texts required urgent responses[12]—some text conversations might take place slowly across days. But even these slow "conversations" now competed with cost-free alternatives as people like Sandra logged onto their MySpace in the evening to see if someone had left a message.

Lastly, there are the more negative communications, including cyber-bullying and other forms of bullying such as crank calls, which are currently being discussed as an issue in other research.[13] However, rather than bullying perhaps a more common process was that of falling out with peers. This sometimes continued with subsequent communications, either to try to sort things out or to communicate more acrimoniously, including some "nasty" texts between ex-boy and -girlfriends. But as was pointed out by several informants, although this may be in the form of texting, the rise of other channels means that such communications is more or equally likely to be online nowadays, either broadcast on IM[14] or as messages left on people's Web sites for third parties to see.

DEALING WITH IMAGES

As in the research on the cameraphone in other countries, this was often the children's first (digital) camera and it was mostly used for capturing spontaneous images from everyday lives such as a group of friends in the park, a new baby, fireworks, or a celebrity. In contrast, those young people who also had digital cameras used them on more special occasions where the taking of pictures was anticipated such as at parties or on holidays—the main reason being the better quality of the images on dedicated devices.

But what counts as amusing pictures for some can count as embarrassing photos for others, although "embarrassing" can take on a range of different meanings.[15] People were embarrassed because of how they looked in a photo quickly taken on the mobile's camera, with both boys and girls noting that this was more of a girl's reaction. Or we can see another example of "embarrassment" in the focus group discussion following.

> *Interviewer:* What counts as a funny photo . . . can you think of any examples from the past?
> *Nina:* Something someone would be embarrassed by (laughs)
> *Sandra:* There's one of me on a trampoline . . . that was really embarrassing . . . no don't (makes a gesture to indicate "don't take a picture")
> *The others laugh*
> *Interviewer (to Sandra):* You didn't go round saying "delete that" . . . it was OK for other people to have the photo?
> *Nina:* I think she tried to get us to delete them . . . but . . . you know (laughs) [clearly they didn't]
> *Sandra:* I was just . . . you know . . . I couldn't be bothered . . . really. They weren't that bad
> *(Focus group 2: 13–14 years)*

Then there were the "annoying" photos, as when James described the practice of being tapped on the shoulder and his peers would take a picture to catch an unguarded expression—a milder version of happy slapping from James's perspective. To be fair, some of the participants said that they generally asked for permission to photograph; they were aware that it was an issue. For others it is not an issue because either the other youth present often posed for pictures or if one was taking pictures of a group the others present also did so. Or else these young people tried to capture images when their peers' guard was down, in the spirit of paparazzi photos, which the young people being photographed may or may not appreciate. Sometimes, when asked to delete a photo, they did it straightaway; sometimes resisting this request itself became a game, a form of teasing.

> *Nina:* This girl fell asleep and she had paint all over her face . . . and we all took photos . . . and shared them around and put them on Bebo . . . and she got very upset . . .
> *Ruth:* And so we took them off.
> *Nina:* When they get upset . . . then you delete them . . . but . . .
> *Several laugh.*

Nina: You keep going until they're upset
(*Focus Group 2:* 13–14 years)

Some, however, were more careful about what pictures they posted on social networking sites.

> *Interviewer:* So, have you ever put any of these pictures or videos [taken on the mobile] on these [Bebo] pages?
> *Mark:* Yeah, but only with my friends' permission (. . .). But I delete anyone that I don't really know now.
> *Interviewer:* You said now. Do you mean this is a new policy?
> *Mark (14):* Yeah, 'cos, I mean, it just suddenly came to me after we had this talk at school about people posting pictures and videos of (other) people who did not want it. I just thought, well, I'll go and delete anyone I don't know and ask (those I do) if I want to put a picture of someone or not of someone on there.

The case of the cameraphone (and video on the camera) raises a number of issues for school. One of the subjects occasionally captured in a photograph or video was teachers themselves, caught unawares (as when one of the teachers was caught dancing "for no apparent reason"). One common subject was school fights, some of which had been posted on YouTube, including mentioning by name the school in question. And there were examples of staged events.

> *Luke (13):* There are like two or three films from the school (online) . . . and you type in '(Name of school)' . . . They were in the drama studio and Justin, one of year 11, he was just sat on a chair . . . and someone came and happy-slapped him right round the face. But he knew they were going to do it. Like he practiced . . . like fake happy-slapping. And they put in on YouTube. It's like still there.
> (*Focus group 2:* 13–14 years)

In fact, during the course of the interviewing an urgent communication was read out to all the classes about such practices since it was felt to be giving the school in question a bad name.

If we turn to what youth actually do with images, as in other research,[16] Multimedia Messaging Service (MMS) is rare mainly because of the cost. A few had sent images, especially if it was part of a tariff package that allowed some free multimedia messages or if there was no free alternative like Bluetooth. The most common practice was to show pictures still on the mobile to friends, then delete many, keeping the "best" ones on the grounds of either quality or content. Transferring images to other people by Bluetooth was less common than transferring music.

MUSIC ON THE MOBILE AND SHARING PRACTICES

One of more relatively recent additions to the mobile phone's functionality at the time of the research was the MP3 player, which for some had become a "must-have" when they upgraded. Unlike the distinction between the cameraphone and digital camera, the participants in this research did not differentiate the mobile's music quality from iPods or separate MP3 players. The superior storage of the latter was its greatest asset. As in the case of cameras, the advantage of the mobile was that you always carried it around anyway—and for the practice outlined below, this was a major plus.

The mobile could also more easily be played aloud for collective consumption as young people took turns to listen to each other's music when going home after school, when traveling together on a coach to sports events, or as when Tim stretched the school rules at break times.

> Most of us, like, we keep our phone so it's on silent or it's off in school and at break time or lunch time we'll take it out and turn it on and listen to music. Tim (15)

Downloading ring-tones had become less popular now that these teenagers could create them themselves, transferring their favorite tunes to the mobile phone or recording them using the "record" function. Showing what music you had was in many ways the equivalent to having a photo album. When socializing and asking to see each other's mobiles, gone are the days when this merely involved looking at its features and checking out texts in some cases. Now friends might work their way through pictures and music on the device—the equivalent, for a previous generation, to having a look at someone's music collection in their homes.

But probably what counts as a new development was the degree and ease of sharing music. In the technology-rich world of these youth there were alternative ways to do this (e.g., sending an attached file of digital music via the Internet). Hence, some teenagers would take music from their own CD collection or from downloads (legitimately bought from sites like iTunes or illegally downloaded) and send it on to friends. However, they could equally well put music on the mobile phone, from whatever source, play it and transfer to by Bluetooth whenever those listening said "Can I have that music?" This created more of a sense of spontaneity, even if to an extent preplanned by virtue of putting music on the mobile, where one could transfer the song there and then as a gift. The mobile phone as storage device enabled such ways of sharing to emerge.

GAMES AND BOREDOM

There was little discussion among young people (or in the research literature) about games on mobiles. Even if they had been on mobile phones from an early stage, were ubiquitous and were played by nearly all the research participants at some stage, these games were only a very small part in their lives. However, it is useful to look at playing practices more closely because the conditions under which youth play provides insight into some uses of the mobile phone more generally.

First we need to consider the specificities of mobile games. Most teenagers played the games that they already had installed on the mobile. Some, like James and Tim, had downloaded games. Others would do so if it was free but otherwise it was not worthwhile enough. That said, the teenagers sometimes talked of missing some of the games that they had had on previous mobiles, such as "Snake." Compared to consoles, everyone recognized most mobile phone games were not so sophisticated, although some had acquired versions of "classics" that first appeared on consoles, such as SIMS 2. However, for various reasons what was on the mobile was good enough for certain purposes. Some, like Carol, had no console (although access to consoles was generally widespread among both boys and girls). Some, like Caitlin, quite liked the easier games and admitted that she was not so good at playing harder console ones, while Nina simply did not find the latter interesting. Sometimes, the time available was limited and so it was useful to have something that could be played quickly.

The other advantage of mobile games is that they were always there. If there was some dead time and with nothing else to hand, the teenagers might play them—long car journeys being a commonly cited example. But if there is nothing really gripping on TV, Tim noted how he would play a game while semi-watching a program because he could not be bothered to go off to get his console. Mobile phone games were low key, convenient, not very demanding, and sometimes slightly more interesting than alternatives.

But it was striking how many times young people of various ages specifically said that they played when they were "bored." It is worthwhile looking at what this means in more detail since children seemed to use the word far more often than many adults. Apart from car journeys, young people could be bored when doing homework and bored when visiting parents' friends—or, as Nina put it, "if you're somewhere where you don't want to be." Using the mobile game was an escape to something more interesting.

Clearly there are quite a few occasions when mobile-phone games were time fillers, but then this was to a lesser extent true of the mobile phone in general. Although the young people explored the phone's functions as soon as they got a new model, they also did so in moments when they were bored. For Sandra, listening to music when bored was an alternative to playing games. And despite the many positive reason for sending texts, several respondents also added that they sometimes sent them because there was nothing interesting to do, or they were not interested in what they were (supposed to be) doing. In these circumstances, not all texts required an answer.

> *Harriet:* It's just . . . if you have a spare moment and you don't know what to do . . . just text them and they don't have to answer back straight away
>
> *Ruth:* And then by the time they text back if you don't feel like talking. . . . . you just don't. (Focus group 2: 13–14 years)

In sum, while earlier the emphasis in the mobile phone literature had been on the use of the mobile phone for socializing in various senses, we also have to be aware of the other moments, often when young people are in some way isolated from the world around them, when the mobile phone's range of media helps young people pass the time.

THE INTERNET AND TV ON THE MOBILE PHONE

We have seen how certain aspects of the mobile phone have found a place in people's lives: as an object of talk, in (albeit evolving) communications practices, and for viewing, sharing, and swapping images and music. These are common, sometimes collective uses embraced with greater and lesser degrees of enthusiasm. Games on the mobile also have a role, although one where use is more individual and solitary. We can reflect now on two more uses of the mobile phone which are not (yet and may never be) quite so established: the use of the Internet and TV on the mobile phone.

Internet access had been possible for a while on many handsets, if we think back to the initial launch of Wireless Application Protocol (WAP). In this research, many of the participants knew they had the option to access the Internet on their phone—in fact, some like Ruth and Sandra complained about the design of the buttons such that it was too easy to hit a button and go online by accident, with the financial cost that this entailed. Like adults, they could critically evaluate when the technologies are fit-for-purpose, noting the small screen or, in some cases, the fact that only some Internet services were offered by operators; the part that, Tim commented, had nothing interesting on it.

> You can go on the house computer and it's like the entire Internet. If you go on your mobile one, it's like a little . . . it's like about a third of what you could do on the normal Internet. Tim (15)

In contrast, the home PC offered a "free" (for these teenagers) and convenient, easy-to-use access to the Internet, which they all thought they would continue to use for most purposes. That said, many had tried the Internet on their mobile phone or talked to others about it and it was quite clear that the chief barrier to use was cost.

> *Izi:* It uses up a lot of your credit.
> *Gareth:* Mine doesn't. It's only if I download.
> *Interviewer:* Why, have you tried it? (To Izi)
> *Izi:* Yeah and all my credit kind of washed out.
> *(Focus group 2: 13–14 years)*

Among the users of the Internet, most were occasional users. Carol had a certain amount of "free" downloading as part of her contract and in general she thought that her access was cheaper than prepay. For her it was quite useful to be able to communicate online away from the surveillance of parents—one of the themes from those who raise concerns about accessing the Internet from this mobile device. Other examples of irregular use were once when the family had been lost while driving and the son Jed came to their aid by looking up Streetmap.co.uk. Several young people, such as Martin, had looked up sports scores (football and cricket), diving into the Internet quickly and getting out again as soon as possible to minimize cost.

Hence, during interviews and focus groups the question was posed as to what they might do online if the price came down, especially if it moved towards the cost of texting, a level of cost they were used to. Although hypothetical, this generated much more interest in the Internet.[17] Some thought they would quite like to simply spend a little more time doing what they already did (e.g., sport-related) without the pressure to get offline quickly. Others could think of some more occasions when it could be useful.

> *Annabel:* My friend, she forgot her homework. So she looked something
> up in Google on her phone and wrote the definition down.
> *Alicia:* Wow. Oh, I want Google (. . .) I'd do my homework on the way
> to school.
> *(Focus group 3: 11–12 years)*

> *Elizabeth:* Well if you had missed the train and you wanted to see what
> the next train time is and the little thingy is closed.
> *Daisy:* Like you can get funny videos and funny pictures on Google and
> stuff and you could show your friends.
> *(Focus Group 5: 11–12 years)*

In addition to such instrumental users, many young people who had become used to increasingly multitasking on the mobile over time, like Luke, considered it might be quite "cool" to go online.

> Yeah, because if you were in class you could go on your Bebo . . . you
> could go on the Internet and do what you liked and no-one would stop
> you. Cos it's so small they wouldn't see it under the table. Luke (13)

And, in the light of what has been discussed previously, it was yet another alternative when bored on the long car journey (or any other such moments).

In fact, they did raise this scenario in relation to TV on the mobile. At this stage very few had seen TV on the mobile or even the relevant adverts and some were surprised by this innovation. At one level there were equivalent discussions to the Internet on the mobile phone, with reservations about the small screen—a full-sized TV would be preferred for most viewing. But some could see moments when it would be useful (as Sandra noted that she might otherwise have had to miss a program episode that she wanted to watch). Apart from that it could be yet another time filler when they were physically mobile. But of course, it depended on the price.

In sum, these two options were not popular (yet). The Internet on the mobile phone was marginal because of affordability and TV was an unknown to all but a few. Unlike many of the other features discussed, they had not become fashionable to have and there were no related peer pressures. However, while it is not guaranteed that these would ever become widespread, at least some young people could see these having a role in their lives, added to their many other media options.

CONCLUSIONS

Looking across the range of multimedia practices, there are three final observations to make. The first is that in such a brief account, one can only indicate the range of use, examples, and some common situations, and this can create the false impression that all youth do everything, homogenizing their experience. Hence, it is worth adding a reminder that even if some practices are more fashionable, the young people in study had different levels of interest and involvement in them, varied levels of technical competencies, and diverse equipment where older or semibroken mobile phones sometimes did not support certain of the options described here.

The second observation is that in various senses one can argue the evolution of these multimedia practices using the mobile phone has been a relatively mundane process, compared to the initial transformations when children first acquired mobile phones and maybe compared to innovations around the initial development of texting.[18] Or perhaps part of that early excitement was on the part of researchers into these developments. Certainly, it was clear these multimedia practices were often seen as being mundane, just part of everyday life, by these young people. But then again, in research on cameraphone images, it has been pointed out how people value the banal,[19] and, indeed, frame their actions, or photos in this case, as being somewhat "ordinary" as part of the very process of participating in their social networks.

The final observation is that since mobile phones have been around they have come to expect new things each year, new things that it was fashionable to have, new activities that one is expected to be able to do: for example, transferring a song to another mobile by Bluetooth because someone else requests it. While these things have become "no big deal" to the youth, if we see them with the social distance of adults removed from this world it is striking how much their technoscape[20] has developed. It is not new to note how increasingly saturated children's lives are becoming with media possibilities, but here we see such processes occurring specifically in relation to the mobile phone. And we must not forget that with this comes a broadening of the range of decisions, a wider range of competencies and maybe even obligations, the development of new norms of behavior, and so on—in general a greater complexity for young people to consider, even if, for the most part, this does not come across as being a major burden to them.

NOTES

1. Specifically on youth and cameraphones, Fausto Colombo and Barbara Scifo, "The Social Shaping of New Mobile Devices among Italian Youth," in *Everyday Innovators, Researching the Role of Users in Shaping ICTs*, ed. Leslie Haddon, Enid Mante, Bartolomeo Sapio, Kari-Hans Kommonen, Leopoldina Fortunati, and Annevi Kant (Dordrect, Netherlands: Springer, 2005), 86–103; Tomoyuki Okada, "Youth Culture and the Shaping of Japanese Mobile Media: Personalization and the Keitai Internet as Multimedia," in *Personal, Portable, Pedestrian, Mobile Phones in Japanese Life*, ed. Mizuko Ito, Daisuke Okabe, and Misa Matsuda (Cambridge, MA: MIT Press, 2005), 41–60; Barbara Scifo, "The Domestication of Camera-Phone and MMS Communication: The Early Experiences of Young Italians," in *A Sense of Place: The Global and the Local in Mobile Communication*, ed. Kristóf Nyíri (Vienna: Passagen Verlag, 2005), 363–74.
2. The research was commissioned by Vodafone and organized through the Digital World Research Centre (DWRC).
3. Jane Vincent, "11–16 Mobile: Examining Mobile Phone and ICT Uses amongst children aged 11 to 16," DWRC Report funded by Vodafone (UK, December 2004).
4. These were recruited by asking for volunteers from the schools and through snowballing amongst the researchers social networks. Recruiting boys to participate in diary studies continued to be a challenge: although they were represented in all age groups there were fewer in each than girls. The reasons given by the boys for not being willing to participate is of interest; we know informally that some boys in the 11–12 age groups but not in this study had lost or had their mobiles stolen. Some parents commented that they were better off without the mobile phone as it was used to bully these younger boys on the school bus. Moreover, since some of these young boys rarely thought to use them anyway, they had not been replaced.
5. A shopping voucher was given in return for the diaries and the interviews and a contribution made to school funds for the focus groups.
6. Leslie Haddon, "The Contribution of Domestication Research to In-Home Computing and Media Consumption," *The Information Society*, 22 (2005): 195–203.

7. Leslie Haddon, "Research Questions for the Evolving Communications Landscape," in *Mobile Communications: Renegotiation of the Social Sphere*, ed. Rich Ling and Per Pedersen (London: Springer, 2005), 7–22.

8. This is also noted in other European countries: European Commission, *Safer Internet for Children: Qualitative Study in 29 European Countries. Summary Report*, May 2007, http://ec.europa.eu/information_society/activities/sip/eurobarometer/index_en.htm.

9. Originally associated with Philippe Ariès, *Centuries of Childhood* (Harmondsworth, UK: Penguin, 1973). For a discussion in relation to ICT, see Leslie Haddon, *Information and Communication Technologies in Everyday Life: A Concise Introduction and Research Guide* (Oxford: Berg, 2004).

10. Rich Ling and Per Helmersen, " 'It Must Be Necessary, It Has to Cover a Need': The Adoption of Mobile Telephony among Pre-adolescents and Adolescents," paper presented at the seminar *Sosiale Konsekvenser av Mobiltelefoni*, organized by Telenor, Oslo, 2000. Available at http://www.telenor.no/fou/program/nomadiske/artikler.shtml.

11. Vincent, "11–16 Mobile."

12. In contrast to Ditte Laursen, see "Please Reply: The Replying Norm in Adolescent SMS Communication," in *The Inside Text*, ed. Richard Harper, Lin Palen, and Alex Taylor (Dordrecht, Netherlands: Springer, 2005), 53–74.

13. Peter Smith, Jess Mahdavi, Manuel Carvalho, and Neil Tippett, *An Investigation into Cyberbullying, Its Forms, Awareness and Impact, and the Relationship between Age, Gender and Cyberbullying, a Report for the Anti-Bullying Alliance* (London: Anti-Bullying Alliance, 2006; http://www.anti-bullyingalliance.org.uk/downloads/pdf/cyberbullyingreportfinal230106_000.pdf).

14. It was reported that because of this practice MSN had been removed from one local library.

15. This becomes salient for interpreting the statistics from surveys asking about embarrassing or unwanted pictures—the answers cover a wide range of situations.

16. Virpi Oksman, "MMS and Its 'Early Adopters' in Finland," in *A Sense of Place. The Global and the Local in Mobile Communication*, ed. Kristóf Nyíri (Vienna: Passagen Verlag, 2005), 349–62; Nicola Döring, Christine Dietmar, Alexandra Hein, and Katharina Hellwig, "Contents, Forms and Functions of Interpersonal Messages in Online and Mobile Communications," *Seeing, Understanding, Learning in the Mobile Age*, proceedings of international conference, April 28–30, Budapest, Hungary 2005; Scifo, "Domestication of Camera-Phone and MMS."

17. Our previous study noted how when WAP first appeared one boy had experimented online precisely because he had free WAP minutes. See Leslie Haddon and Jane Vincent, "Managing a Communications Repertoire: Mobile vs Landline," paper for the 5th Wireless World Conference, Managing Wireless Communications, 15–16 July, http://members.aol.com/leshaddon/Wireless5.html.

18. Richard Harper, Alex Taylor, and Lin Palen, eds., *The Inside Text: Social Perspectives on SMS in the Mobile Age* (Dordrecht, Netherlands: Springer, 2005).

19. Ilpo Koskinen, "Managing Banality in Mobile Multimedia," in *The Social Construction and Usage of Communication Technologies: European and Asian Experiences*, ed. Raul Pertierra (Singapore: Singapore University Press, 2007), 48–60.

20. Christian Licoppe, "Connected Presence: The Emergence of a New Repertoire for Managing Social Relationship in a Changing Communication Technospace," *Environment and Planning D: Society and Space*, 22 (2004): 135–56.

5   Mobile Communication and Teen Emancipation

Rich Ling

INTRODUCTION: ADOLESCENCE AND MOBILE COMMUNICATION

Technology adoption and use are in flux. In addition, adolescence is a de facto period of flux. Teens use technologies for the purposes at hand. These change and vary according to the point in the emancipation process in which they find themselves. Technology is also configuring the way that adolescents, and their parents, are carrying out the emancipation process. Starting in the 1990s, teens first recognized the potential for the use of wireless communication. Beginning with the ownership and use of pagers, teens started to use these devices for various forms of coordination and interaction. They have also been active in exploring innovative forms of use.

Statistics from the period show that in Norway in 1997 only about 20 percent of school-aged teens owned a mobile telephone. If we move two years further along, two-thirds of this same age group owns a mobile phone. In the period between 1997 and 1999 we saw the introduction of prepaid subscriptions, subsidized handsets, and teens' "discovery" of Short Message Service (SMS) as, at that time, a free channel of interaction. These elements, along with the intense desire to communicate and the need to establish themselves as independent actors, led teens to adopt the mobile phone.

These statistics suggest that there was a simple and largely unproblematic adoption of a new technology. This is, however, not the case. The adoption process—at least as observed in Norway—was not always an easy one. It challenged existing social conventions, ideas regarding children, and their place in the communication hierarchy. Thus, the interactions between parent and child were sometimes long and bitter with regard to the eventual purchase of a mobile communication device.

In a series of interviews done in the spring of 1997, we were able to pick up the strains occasioned by teens' desire for the ownership of communications technology. In one case the discussion centered on the purchase of a pager by a sixteen-year-old respondent.

> *Kristine (mother):* There has been continual discussion about both mobile telephones and pagers and we have said no.

Frank (son): I have ordered a pager now.

Interviewer: Why does your mother say no to a pager? Is it the cost or is it . . .

Kristine: Yeah, there is that and of course I don't think that he is so important that he needs to be reached right then.

Frank: But a pager is not a businessman thing any more. There were two pages in the newspaper about how all the teens have pagers.

Kristine: Yeah, but when you are at school you are at school and when you are at home you are at home.

Frank: Yeah, but I do not have any plans for having a pager at school.

Kristine: Then they can get in touch with you here at home. I think it is a bad thing with mobile telephones on the bus and at school. If you are at home then you can be reached there, if you are not home then ok, it is not so important that they can wait to call. They can call later.

The citation shows that the lines of argument were well rehearsed between these two. The presence of the external interviewer with the recording equipment meant perhaps that the comments were uttered in a civil tone, but only just. The interaction was, however, one of engaged principle. It is easy to see that the mother and the son could easily have had a far more heated discussion.

On the one hand, Frank has been so brave as to order a pager against the will of his mother. Frank—who was in the middle of his adolescence, at least after Norwegian standards—had, in effect, marked his own independence through the purchase of what we now see as an antiquated device but was at that time a common part of teens' kit. Kristine, on the other hand, had obviously developed her arguments against the device. The first line of defense was that only certain people in society were so important—or perhaps pompous—so as to need this type of communication device. Part of the subtext here is that Frank is not so in demand so as to need a pager. His radius of action is limited to a small number of locations where either he does not need to be reached or he is available by other means.

Frank and Kristine were involved in a negotiation regarding the meaning of mobile communication devices. At that point, they were not taken for granted as they now are. There was both a discussion about the instrumental necessity and also the morality of the device and what constituted a legitimate need. The various actors fall on different sides of the issue and develop their arguments accordingly. This time capsule is interesting in that it shows how the meaning of the device was being negotiated between teens and their parents from its very introduction.

Viewed from the distance of a decade, these discussions seem odd. Now all Norwegian teens—and many preteens—have mobile phones.

As of 2005, there were quite literally no teens in Norway that did not have a mobile phone.[1] Adolescent adoption and use of mobile communication, for example, has affected the way that mobile technology has developed. Their wholehearted embracing of SMS has led to the development of various input devices and forms of argot. In addition, it has led to changes in the design (the color, form, and "flashiness") of mobile handsets. Given this situation, there is not a discussion as to whether, but *when*, a teen will receive a mobile phone. The mobile telephone is now seen as an integral part of teenage life in Norway. The fact that all teens in Norway own one means that, in some respects, mobile phones are not remarkable anymore.

THE EMANCIPATION OF TEENS AS A CRITICAL SOCIAL JUNCTURE

Leaving the Family of Orientation

The introduction of mobile communication devices has affected the lives of teens. It has seemingly rearranged the way that teens move in the world and the way that they constitute themselves as independent actors in society. The device jiggles things just enough so as to cause people to think about the true nature of adolescence and parenting. The analysis of adolescence and the corresponding analysis of emancipation are well-established academic genre.[2] In addition to historical studies,[3] Hugh Cunningham[4] examines the general situation of childhood and youth, as did James Coleman[5] and John Demos and Virginia Demos.[6] There is a rich literature on the sociolinguistics of adolescence,[7] as well as juvenile delinquency.[8] There is also a well-developed line of research examining the impact of media on children and the so-called new media.[9]

These analyses are done, of course, against the backdrop of adolescence, that odd phase of life that is a combination of leaving and becoming. This is played out in different ways, at different times, and at a different tempo in each culture. However, the individual leaves the confines of the family sphere and, as time passes, becomes reestablished as a more or less self-sufficient individual in a new (or at least a slightly rearranged) set of feathers. Indeed, emancipation is one of the central processes associated with adolescence. Learning the ability to function outside the family sphere is a central project for this period of life. In Norway, as in many other places, the teen must master the skills of managing personal economy, strategies for negotiation within various institutions, interactions with others, the role of sex in one's life, how one secures a job, the expectations of the working world, and a sense of personal style and integrity. The nascent adult must learn to take responsibility for his or her own actions.

Emancipation as a Process

The idea of emancipation might conjure the idea of a gradual transition from one status to another. The emancipation process, however, is not a "onetime" procedure. Rather, it is a whole series of episodes and trials that the adolescent and parents confront. As we saw in the interaction between Frank and his mother, the purchase of a pager was an event that tested the will of both partners. The needs and interests of ten-year-olds are not the same as those of their eighteen-year-old brothers or sisters. While, at least in the Norwegian context, the former are firmly entrenched in the sphere of their parent's home, the latter are often en route to their own separate identity. The former have a limited geographical field in which they move and they are often dependent on their parents for money. Along with the physical changes of puberty, the older teens often have a wider geographic range, they may earn their own money, and may be focused on a career or a job path. Finally, many persons this age have also confronted worldly issues such as alcohol use and sexuality. This picture is clearly drawn in broad terms. Many younger children match their older siblings when thinking of these issues and the opposite is also true. Nonetheless, ten- and eighteen-year-olds quite often stand on opposite sides of the adolescence divide. Barring perhaps retirement, this period sees the most dramatic changes we experience in our lives.

Emancipation as a Consequence of Industrialization

Far from being a simple marking of physical maturation, adolescence is a period in which children gain the skills they need to carry them through the subsequent portions of their lives. This is a complex process in industrialized, postindustrial, and now information-based society.

In traditional society, children could learn at the feet of their parents or a master of some trade.[10] However, in posttraditional society, the child simply replicates the education of and socialization of his or her parents. The professions and life experiences of those who came before may not adequately reflect the reality of the child. The changes in technology and in the structures of society mean that the experiences of a previous generation are, to some degree, outdated. The child must discount the experiences of the adults. Thus, unlike in traditional societies, the adolescent is active in the formation of his or her transition from childhood to adolescence and even more so in the transition from adolescence to adulthood.[11] Looking at the broader situation of society, there is a need to develop laborers who are calibrated for the challenges to be met in the current era, not the challenges that were relevant for an earlier generation. Significant portions of socialization are removed from the home and entrusted to the school system. According to work done by Jeffery Arnett, people in the United States consider the transition from adolescence into adulthood completed when the individual is, for example, capable of accepting personal responsibility for his or her own actions, establishing his or her own sets of beliefs and values, operating on an equal footing with parents, and being financially independent.[12]

The Peer Group as a Central Link in Emancipation

The same-age peer is a central agent of socialization in the transition from childhood to adulthood. Indeed, this is an important link to understand when thinking about the role of the mobile telephone and other mediated forms of interaction. This group takes a central position in the child/adolescent's activities, his or her sense of identity, consumption patterns, and in his or her orientation.[13] Moreover, peer culture is a major element in the process of developing notions of personal identity. It is a sphere where the individual functions as a colleague and partner.[14] Participation in the inner life of the peer group is also a motivation for the adoption of communication technology.[15] The peer group allows adolescents to develop their social skills in a relatively protected group. Peers help the adolescent to experience self-esteem, reciprocal self-disclosure, emotional support, advice, and information. They give the child experience in dealing with in- and out-groups, and in this way they also give the individual experience with intrigues, disagreements, power plays in miniature, the ability to vulnerable and various types of infighting.[16] In all of this, communication, lots of communication, is central.

These groups are largely protective of their members. They draw a symbolic and perhaps ideological boundary around themselves and resist the intrusion of others. This is seen in the development of what Gary Fine calls "idioculture" and that may include a whole system of nicknames, jokes, styles of clothing, songs, artifacts, attitudes, political orientation, and so on.[17] While there is support in the peer group, there is also teasing, gossip, and infighting. While it has profound influence on the selection of certain cultural items such as slang and clothing, parents and the adult world are influential in areas such as career choices.[18] Fine has indicated that these activities may indeed provide the basis for enhanced group solidarity and loyalty.[19]

TEENS' ADOPTION OF MOBILE PHONES

As mentioned, the notion of adolescence suggests that the experience of one generation is necessarily different from that of their parents and grandparents. This implies that they need to have different skills and that the way that they experience and develop their cultural practices is different form previous cohorts.

Mobile Communication and Teen-Parental Interaction

In many countries, the mobile phone worked into the way that teens and parents achieve emancipation. It has meant that parental concerns regarding the safety, security, and control of their children have been carried out differently. Prior to the mobile phone, teens often sought the luxury of a telephone extension in their own room. Rather than having access to a telephone in a fast location, the mobile phone allows the teen both mobility and accessibility. It has given teens access to the world outside the parental sphere in a way that has not been experienced by earlier cohorts. According to Cunningham, "children have demanded and received earlier access to the adult world; they have not been willing to accept the attempt to prolong childhood into the late teenage years."[20] The mobile phone provides cautious autonomy and serves as a type of extended umbilical cord or ("Hi, Dad, it is raining and I am freezing; can you come and get me?").[21]

When thinking about the function of the device within the family context, it has facilitated family coordination ("Come home to dinner, NOW!"). Nicola Döring and her colleagues found that quite often parents were teens' most common interlocutors.[22] These calls were often associated with organizing transportation between activities.

The mobile phone plays into the interaction between parent and child. It gives the child freedom from his or her parents that were unheard of a decade prior. On the other hand, it is a type of umbilical cord and a medium through which parents can maintain contact with—and exercise their power over—their child. While the child can use the device to operate more freely, the parents can also impose a new form of economic fettering in that the child has to pay for their use of the phone.

The Mobile Phone and Peer Interaction

While the mobile phone has had meaning for the interaction between parent and child, it has truly found its role in facilitating how teens interact among themselves. The mobile phone has meant that the adolescent has closer contact with the peer group.[23] It has facilitated interaction between peers at many levels. It facilitates instrumental interaction (coordination) as well as various forms of expressive use (phatic interaction, the expression of style through the ownership of different handset types, the inclusion/exclusion in peer groups through the use of argot, status position via the number of calls received or names in the name register, and so on). In some cases it is also used as a platform for the use of advanced services.

INSTRUMENTAL USE OF THE MOBILE PHONE

The mobile phone gives teens control over their own communication channel. This means that when they call to a friend they know that the friend and not their parent will answer the phone. In addition, they know that when that friend calls or sends a text message, the call will come directly to the interlocutor. This is a type of control over the communication process that was not possible with traditional landline telephony.[24] As with other groups, the mobile phone is used for instrumental purposes such as safety/security and coordination. Considering safety, the mobile phone is used as a way to contact friends when trouble arises. The emergencies might simply be the need to have a good friend available or in the other extreme—and somewhat perverse version of "safety"—it can be in the form of organizing the gang for confrontations with others.[25]

The impact of the device is also seen in the way that teens plan their activities. Coordination takes place in real time.[26] This type of coordination can be seen in mundane events of synchronizing interaction. As an example, I was recently teaching a large class and I asked the students whether they had received any text messages since they had been in the lecture hall. Among the many that raised their hands, two women said that they had planned to meet in the class, but when, upon arriving, they could not see one another they simply sent text messages and were able to find each other. This is a seemingly minor event in their daily life. It is indeed a minor event in the broad sweep of their daily interaction. Nonetheless, it underscores an important dimension of mobile, and individually addressable, communication. It is what I have referred to as microcoordination.[27] Namely, we can contact individuals regardless of where they are and coordinate mundane interaction.

EXPRESSIVE USE

As noted earlier, the mobile phone, the actual physical object, has become an icon of teens in the last decade. Along with other types of media, the mobile telephone has served to demarcate the generations and to "detach . . . young people from their parents, their class and their locale."[28] In addition to its instrumental functions, the mobile phone, as with other types of media devices, can be an element in the formation of identity for the teen.[29] The brand and the model of the handset can say much about the owner.

The popularity of a particular brand or model will wax and wane. However, the use of the device to mark status and to mark membership in a group is a part of the appeal of mobile phones.[30] Devices with different facades and different functionality have been designed for and adopted by different groups. Youth telephones have included features such as MP3 players, cameras, flashy colors, and various gaming functions. In this way, the mobile phone is seen as a type of fashion item, in addition to its role as a tool for communication. As with other fashion items, however, the codex of what is "in" continually shifts.[31] The mobile phone becomes a part of the teens' socialization into the management of personal display. By contrast, those who use other types of handsets (soccer mom handsets, business guy handsets, hopelessly old "museum pieces") and those who are not in tune to the lingo mark their position outside of the circle.[32]

Yet another aspect of mobile communication is that it provides teens with a forum in which they can establish their own form of slang and argot. As with the ownership of particular handsets, the use of argot is a way with which to mark inclusion in the group. This can take the form of using special words or special formulations. Examples of this include, for example, the replacement of words with single letters where possible and the use of ellipsis as a marker of a more vocal style of interaction.[33] While the use of these formulations is much commented in the media, their actual adoption is often far less comprehensive. The exact mixture can vary from culture to culture,[34] but the point is that the use of slang and argot is a way for teens to identify with the group in their interactions.

THE MOBILE PHONE AND THE EXPLORATION OF DIFFERENT SIDES OF LIFE

Another, less savory, side of mobile communication among teens is that it allows room for coarser forms of interaction. There is the issue of very personalized mobile bullying/chicanery, and the exchange of nude and pornographic images via Multimedia Messaging Service (MMS) is a flourishing practice in some groups.[35]

The camera function of the mobile phone means that we are all perpetually carrying a camera with us. Paparazzi photos and illicit photos of noncelebrities in unguarded moments are a new dimension of social interaction.[36] In addition, the ability to exchange digital images and ill-judged text means that photos and textual interaction are not simply lost in the ether at the end of the interaction. Among those teens who are extremely heavy users of the mobile telephone, there is a greater incidence of offences such as drinking, theft, narcotics use, fighting, and other illicit forms of behavior.[37]

CONCLUSION: THE MOBILE PHONE AS A TOOL IN THE EMANCIPATION OF TEENS

The analysis here has examined the different ways that the mobile phone has effected the emancipation process for teens in Norway. The device has created a space in which the teen can develop his or her own routines for interaction with peers. It has become a type of icon for teens in the sense that the actual form of the device is important. The type of mobile phone owned by an individual is important and is interpreted by peers in their total estimation of the individual. The device allows for various forms of status estimation via the number of names in the name register, the number of calls and text messages being received, and so on. Mobile telephony has facilitated a type of perpetual interaction between members of a peer group, and indeed the mobile telephone is often seen as a device that has facilitated group cohesion.

Transition from adolescence to adulthood is one of a certain separation between parents and child where the child increasingly—and perhaps falteringly at times—creates his or her own space for action. The creation of the space is, to some degree—and again increasingly—dependent on links to peers. Within this personal space, the teen must explore the various issues associated with taking responsibility for his or her own actions. These can include the realms of personal economy, education, health related (the role of sexuality, drugs, alcohol, etc.), and the organization of a personal style. The individual needs to establish some sort of value system that may well include bits and pieces form his or her parents' or peers' values; or it might be a total rebuilding of those values into something that the nascent adult feels better fits his or her life situation.

Mobile communication has clear impacts on the transition from child to adolescent and finally to adulthood. The mobile phone provides the link between the teen and his or her peer group. Thus, the device can help to open up a personal space for teens in which they can begin to explore the different issues at hand. It provides for an active link to others who are also exploring the same issues and making decisions as how they can tackle them. In the same device, there is a coordination tool and an artifact with which they can express their budding sense of style and identity. It is a device through which they can examine the different issues of the day with their peers and, perhaps in concert or perhaps individually, patch together solutions to these issues in their movement towards adulthood.

NOTES

1. Odd Vage, *Mediabruksundersøkelse* (Oslo: Statistics Norway, 2006).
2. In English we speak of emancipation, and in Norwegian the phrase is *løsrivelse* or disconnection from the family. This is a much more limited use of the word than, for example, Habermas's notion that implies a separation from social tradition or the conservative forces in society. See, for example, Craig J. Calhoun, *Habermas and the Public Sphere* (Boston: MIT Press, 1992).
3. Philipe Ariés, *Centuries of Childhood* (New York: Random House, 1960).
4. Hugh Cunningham, *Children and Childhood in Western Society since 1500* (New York: Longman, 2005).
5. James Coleman, *The Adolescent Society: The Social Life of the Teenager and Its Impact on Education* (Westport, CN: Greenwood Press, 1961).
6. John Demos and Virginia Demos, "Adolescence in Historical Perspective," *Journal of Marriage and the Family*, 31 (1969): 632–63.
7. See, for example, Angie Williams and Crispin Thurlow, *Talking Adolescence: Perspectives on Communication in the Teenage Years* (New York: Peter Lang, 2005).
8. Delbert Elliott, David Huzsinga, and Susan Ageton, *Explaining Delinquency and Drug Use* (New York: Sage, 1985).
9. Kristin Dortner, "Difference and Diversity: Trends in Young Danes' Media Use," *Media, Culture and Society* 22 (2000): 149–166; Kristin Dortner, *Unge medier og modernitet—pejlinger i et foranderligh landskab* (Valby, Danmark: Borgens forlag 1999); Rich Ling and Per Helmersen, " 'It Must

Be Necessary, It Has to Cover a Need': The Adoption of Mobile Telephony among Pre-adolescents and Adolescents," in *The Social Consequences of Mobile Telephony*, ed. Rich Ling (Oslo: Telenor 2000); Rich Ling and Leslie Haddon, "Mobile Emancipation: Children, Youth and the Mobile Telephone," in *International Handbook of Children, Media and Culture*, ed. Kristin Dortner and Sonia Livingstone (London: Sage, forthcoming); Sonia Livingstone, *Young People and New Media* (London: Sage, 2002).

10. Philipe Ariès, *Centuries of Childhood*.

11. Barney Glaser and Anslem Strauss, *Status Passage* (London: Routledge & Kegan Paul, 1971), 57–88.

12. Jeffery Arnett, "Adolescents' Uses of Media for Self-Socialization," in special issue on "Adolescents' Uses of the Media," *Journal of Youth and Adolescence*, 24 (1995): 519–33.

13. Dennis P. Hogan, "Parental Influences on the Timing of Early Life Transitions," *Current Perspectives on Aging and Lifecycle*, 1 (1985): 1–59; Aenaida Ravanera and Fernando Rajulton, *Parental Influences on the Timing of Early Life Transitions: Evidence from the 2001 General Social Survey* (London, Ontario: Population Studies Centre, University of Western Ontario, 2004).

14. Rich Ling and Birgitte Yttri, "Hyper-Coordination via Mobile Phones in Norway," in *Perpetual Contact: Mobile Communication, Private Talk, Public Performance*, ed. James Katz and Mark Aakhus (Cambridge: Cambridge University Press, 2002), 139–69.

15. Ryan Adams and Brett Laursen, "The Organization and Dynamics of Adolescent Conflict with Parents and Friends," *Journal of Marriage and Family*, 63 (2001): 97–110; Peggy Giordano, "The Wider Circle of Friends in Adolescence," *American Journal of Sociology*, 101 (1995): 661–97; Susan Harter, "Self and Identity Development," in *At the Threshold: The Developing Adolescent*, ed. Shirley Feldman and Glen Elliott (Cambridge, MA: Harvard University Press, 1990), 352–87; Allison Ryan, "Peer Groups as a Context for the Socialization of Adolescents' Motivation, Engagement, and Achievement in School," *Educational Psychologist*, 35 (2000): 101–11; Ritch Savin-Williams and Thomas Berndt, "Friendship and Peer Relations," in *At the Threshold: The Developing Adolescent*, 277–307; James Youniss and Jacqueline Smollar, *Adolescent Relations with Mothers, Fathers and Friends* (Chicago: University of Chicago Press, 1985).

16. William Corsaro and Donna Eder, "Children's Peer Cultures," *Annual Review of Sociology*, 16 (1990): 197–220; Peggy Giordano, "Relationships in Adolescence," *Annual Review of Sociology*, 29 (2003): 257–82.

17. Gary Fine, *With the Boys: Little League Baseball and Preadolescent Culture* (Chicago: University of Chicago Press, 1987).

18. Clay Brittain, "Adolescent Choices and Parent-Peer Cross Pressure," *American Sociological Review*, 28 (1963): 385–91.

19. Gary Fine, *With the Boys*.

20. Hugh Cunningham, *Children and Childhood*.

21. Manuel Castells, Mireia Fernandez-Ardevol, Jack Qiu, and Araba Sey, *Mobile Communication and Society: A Global Perspective* (Cambridge, MA: MIT Press, 2007); Rich Ling and Leslie Haddon, "Mobile Emancipation: Children, Youth and the Mobile Telephone," in *International Handbook of Children, Media and Culture*, ed. Kristin Dortner and Sonia Livingstone (London: Sage, 2008).

22. Nicola Döring, Katharina Hellwig, and Paul Klimsa, "Mobile Communication among German Youth," in *A Sense of Place: The Global and the Local in Mobile Communication*, ed. Kristóf Nyíri (Vienna: Passagen Verlag, 2005), 209–17; Rich Ling, "'We Release Them Little by Little': Maturation and

Gender Identity as Seen in the Use of Mobile Telephone," paper presented at the *International Symposium on Technology and Society (ISTAS '99), Women and Technology: Historical, Societal and Professional Perspectives* (New Brunswick, NJ: Rutgers University Press, 1999).

23. Mizuko Ito, "Mobile Phones, Japanese Youth and the Re-placement of Social Contact," in *Society for the Social Studies of Science* (Boston: Society for the Social Studies of Science, 2001); Christian Licoppe, "Connected Presence: The Emergence of a New Repertoire for Managing Social Relationships in a Changing Communications Technoscape," *Environment and Planning*, 22 (2004): 135–56; Rich Ling, "'It rings All the Time': The Use of the Telephone by Norwegian Adolescents," *Report for Telenor Forskningsinstitutt* (Kjeller: Telenor R&D, 1998); Rich Ling, "'It Is "in": It Doesn't Matter if You Need It or Not, Just that You Have It,'" in *Machines that Become Us*, eds. James Katz (New Brunswick, NJ: Transaction, 2003).

24. Lydia Sciriha, "Teenagers and Mobile Phones in Malta: A Sociolinguistic Profile," in *New Technologies in Global Societies*, ed. Patrick Law, Leopoldina Fortunati, and Shanhua Yang (London: World Scientific, 2006), 159–78.

25. Siv Håberg, *Tilgjengelighet til glede og besvær: En studie av bruk av holdninger til mobiltelefon som ny teknologi* (Institutt for Kulturstudier, Universitetet i Oslo, Oslo 1997).

26. Rich Ling and Birgitte Yttri, "Hyper-coordination via mobile phones in Norway."

27. Rich Ling and Birgitte Yttri, "Hyper-Coordination via Mobile Phones in Norway."

28. Sonia Livingstone, *Young People and New Media: Childhood and the Changing Media Environment* (London: Sage, 2002).

29. Jeffery Arnett, "Adolescents' Uses of Media for Self-Socialization," *Journal of Youth and Adolescence*, 24 (1995): 519–33.

30. Leopoldina Fortunati, "Mobile Telephone and the Presentation of Self," in *Mobile Communications: Re-negotiation of the Social Sphere*, ed. Rich Ling and Per Pedersen (London: Springer, 2005), 203–18; Leopoldina Fortunati, James Katz, and Raimonda Riccini, eds., *Mediating the Human Body: Technology, Communication and Fashion* (London: Lawrence Erlbaum Associates, 2003); James Katz and Satomi Sugiyama, "Mobile Phones and Fashion Statements: Evidence from Student Surveys in the US and Japan," in *Mobile Communications: Re-negotiation of the Social Sphere*, ed. Rich Ling and Per Pedersen (London: Springer, 2005), 63–81.

31. Rich Ling, "It Is 'In.'"

32. Anita Lynne, *Nyansens makt—en studie av ungdom, identitet og klær* (Lysaker: Statens Institutt for Forbruksforskning, 2000).

33. Rich Ling and Naomi Baron, "The Mechanics of Text Messaging and Instant Messaging among American College Students," *Journal of Sociolinguistics*, 26 (2007): 291–98.

34. Bella Ellwood-Clayton, "Desire and Loathing in the Cyber Philippines," in *The Inside Text: Social Perspectives on SMS in the Mobile Age*, ed. Richard Harper, Alex Taylor, and Leysia Palen. (London: Klewer, 2005), 195–222.

35. Petter Brandtzæg, "Children's Use of Communications Technologies," in *Mobile Media, Mobile Youth*, ed. Gitte Stald (Copenhagen: University of Copenhagen, 2005).

36. Tarja Tikkanen and Amund Junge, *Realisering av en visjon om et mobbefritt oppvekstmiljø for barn og unge* (Rogalandsforskning, Stavanger, 2004); Marion Campbell, "The Impact of the Mobile Phone on Young People's Social Life," in *Social Change in the 21st Century* (Brisbane: Queensland

University of Technology, 2005); Helmut Kury, Silvia Chouaf, Joachim Obergfell-Fuchs, and Gunda Woessner, "The Scope of Sexual Victimization in Germany," *Journal of Interpersonal Violence,* 19 (2004): 589–602.
37. Rich Ling, "Mobile Communications vis-à-vis Teen Emancipation, Peer Group Integration and Deviance," in *The Inside Text: Social Perspectives on SMS in the Mobile Age*, ed. Richard Harper, Alex Taylor, and Leysia Palen (London: Klewer, 2005).

6   Mobile Media and the Transformation of Family

Misa Matsuda

In this chapter, I will focus on the transformation of the Japanese family—in particular how parent-child relationships are mediated by the use of mobile media. Some may question how this topic—which focuses specifically on mobile media in Japan—is relevant to a global audience. It is important to recognize that the rise of global mobile media is framed by the forces of the local. To gain acuity on the trends of mobile media adoption and adaptation we must understand that technologies do not exist in a vacuum and that the process of technological adoption is part of a dynamic social process. Through the lens of the *keitai*[1] we can gain insight into the complex and dynamic relationship between technology and society.

WHY FOCUS ON JAPAN?

Since the latter half of the 1990s, mobile phone trends in Japan have been attracting global attention. Initially, mobile phones in their miniaturized form were merely the latest example of innovative technology in the country renowned for its global production of commodities such as automobiles, transistor radios, and Walkmans.[2] With the rapid spread of mobile Internet services (Internet use from mobile phones) from 1999 onwards, international attention turned to Japan as a model for the potential future of mobile media; the mobile phone as a multimedia tool par excellence.[3]

Subsequently, Japan's mobile phones—or *keitai* as they are called—boasted sophisticated multimodality that encompassed the digital camera, GPS, and music download capabilities. However, in order to understand the rise of this technology in Japan, it is important to outline the very specific sociocultural factors informing the uptake. In short, one needs to understand the rise of *keitai* culture—emblematic of innovative global mobile media—as a socio-technology. As I have argued elsewhere, it is imperative that the specific circumstances that underlie the national trend of the *keitai* be contextualized.[4]

The most distinctive aspect of the use of *keitai* is the preexisting use of the pager, a trend that was at its peak among Japanese youth in the mid-1990s. The pager was first established as a mobile media gadget among the younger generation. The centrality of youth in the adoption of mobile media led to the development of the core uses of the *keitai* Internet—wallpaper and ring-tone downloading. Moreover, the text message exchanges that then became common practice among Japanese youth led to the use of mobile Internet e-mail that is ubiquitous among all age groups today.

The second underlying condition of the mobile Internet unique to Japan is the fact that in the late 1990s Internet use was not as widespread in comparison to other technologically advanced nations. I have previously cited many reasons for this,[5] but the important issue is that there had not been much advancement in the spread of the PC—the device necessary as a precursor to Internet usage. Amidst such circumstances, the *keitai* was established as a cheap, user-friendly Internet terminal around 2000, leading to the rapid diffusion of the mobile Internet.

The third unique factor is that due to industry circumstances, Japan did not employ Global System for Mobile (GSM) networks in second-generation mobile phones. Although text messaging—known as short message services (SMS)—became popular among Scandinavian teenagers in the latter half of the 1990s,[6] different operating system in Japan of text messaging via *keitai* that began in 1996 meant that these messages were only possible between subscribers of the same service providers. This incompatibility made short message services unpopular in Japan and led to the eventual usage of the more universally accessible e-mail systems.

More fundamental to the unique circumstances of the *keitai* in Japan, this third factor is strongly connected to the adoption of a business model referred to as the "vertically integrated model." This model of the *keitai* industry is characterized by the position of leadership taken by the telecommunications carrier in determining the specifications of the handsets and services, and where the manufacturers and the content providers comply with those criteria. *Keitai* handsets are developed in accordance with the new services proposed by the telecommunications carrier. Through incentives (sales promotion premiums) paid to the distributors, costs can be delivered to the users at a rate far below the initial price. The popularity of the mobile Internet and the multifunctional high-performance handset was driven by such industry, rather than users' needs.

Due to these particular factors informing the rise of the *keitai*, this model may not be as useful as one might expect in understanding the future of a homogenous rise of global mobile media. Rather, just as the uptake of mobile-phone practices were informed by the local, so too mobile media demonstrates that global convergence is marked by divergence. Similarly, in this chapter, the topic of mobile media is placed in context of the unique circumstances of Japan as a culture or society in order to stimulate awareness that *keitai* use is embedded in the lifestyles and family relationships inherent in each culture or society.

Moreover, through this case study, there will undoubtedly arise examples that allow for parallels between Japan and other contemporary postindustrial societies. I also expect that the particular underlining traits of mobile-media-embedded contemporary societies will also become clearer after examining the situation in Japan.

TOWARD THE *KEITAI*-MANAGED FAMILY

Mother of eight-year-old and ten-year-old sons (on her way to pick them up from their after-school lessons),

> I always call to say, "We're done, so we're coming home." Once, when I neglected to call, my husband scolded me saying, "Since you have a *keitai*, you could at least call!"
>
> *Author:* But doesn't the lesson always end at the same fixed time?
> *Mother:* Yes, but even still, I always call home to announce that we're on our way. My husband also calls from the office without fail.[7]

Since my research of the *keitai* from 1995 I have noted one particular phenomenon that has emerged—the *keitai*'s effect on interpersonal relations: "the increase in individual selectivity."[8] Though the *keitai* can, in theory, be used anytime anywhere, people are actually utilizing the *keitai* to foster already existing relationships and social capital. Similarly, this trend can also be applied to *keitai*-embedded familial relationships.

Let us first examine the transformation of the Japanese family itself. Yoriko Meguro characterizes changes in the Japanese family after World War II as a shift towards "individualization." Prior assumptions that family is a natural outcome of being together or sharing blood ties have been displaced by this notion of family as something to be created and maintained through the ongoing efforts of individuals.[9] Following on Meguro's work, others have argued that certain changes to the Japanese family can be understood by terms such as "individuation" and "privatization" as well as by "individualization."[10] Such a tendency is not a distinctive characteristic of the Japanese family. Rather, it is in line with the trends of familial relationships in Europe and America.[11] The same can also be said of the national concern for "the crisis of family"; it is a global societal epidemic integrally linked to "individualization."[12]

However, within this global phenomenon of arising individualization, there are also great differences. One key distinction between Japanese from European and American families is the strong tendency for Japanese to get married legally.[13] In spite of the fact that it has become "the choice of the individual to lead or deny a family-like lifestyle,"[14] people believe in a decision to "create" a family, at least in sentiment. Under these circumstances, what is valued among family members is intimacy, or more concretely, communication through words. While there is a steady rise in the number of families who have spoken dialogue,[15] the "tendency to desire increased conversation within the family" has not waned. As aforementioned, "the crisis of family" is a common topic of discussion in contemporary Japan. But in actuality, there is in fact a strengthening and greater implementation of values toward "family centrism" in Japanese society.

Hand in hand with the trend towards the "individuation" of daily life, however, opportunities for copresence and communication have been decreasing.[16] As a result, the family now makes optimal use of the *keitai* in order to maintain their familial bonds. In addition to "necessary" calls home there are also "routine" *kaeru* (Japanese for "I'm coming home") calls that are made out of consideration to the family member who is at home waiting.

The *kaeru*-call was popularized by a 1985 campaign by the newly privatized NTT national telephone carrier to promote telephone calls. Now the *kaeru*-call has caught on and the term is commonly used to refer to "calls made to the home to inform the family of the time of one's return." As noted earlier in this part's opening quote, the mother said, "Since my eldest child will soon be going to after-school lessons on his own, I intend to have him carry a *keitai*." When alone, that child is certain to make *kaeru*-calls home from his *keitai* since this is his family's standard practice.

Furthermore, when each member of the family is busy, he or she can use a *keitai* to communicate and adjust schedules in an effort to create copresence. For example, a family dinner can be arranged because of the *keitai*. Similarly, after quarreling with a family member, while it may be difficult to give voice to a simple apology such as "I guess I went a little too far," the *keitai* allows one to express these words more readily. Text messages to express something that is difficult to say in person are also sent between married couples as well as between parent and child.

While not used among families who believe that "because we are family, there is understanding without dialogue," *keitai* are used in families who believe that "because we are family, we must speak to one another," and serves as a tool that facilitates intimacy. In the midst of a strengthening and implementation of values toward "family centrism," the *keitai* has become a "binding agent" that maintains the intimacy of the family unit.

PARENT SURVEILLANCE VIA *KEITAI*

A mother of seven-year-old and ten-year-old daughters avows, "My daughter said to me, 'Wouldn't you also feel better if I carried a *keitai*?'"

According to a survey by the Mobile Society Research Institute conducted in March 2005, the rate of *keitai* ownership of upper-grade elementary school students (fourth to sixth graders) is 24.1 percent in metropolitan areas. For junior high school students it is 66.7 percent and 95.0 percent for high school students. A child's induction to cram school or after-school lessons is the most common response (42.5%) to the question on what prompts upper-grade elementary students to carry a *keitai*. And the percentage of those who used a *keitai* because of the prevalence of friends who use them—the most common motive for those in junior high school and above—is only 2.5 percent.[17] The data on the status of *keitai* use show that their primary use—at least for elementary school children—is for communication with the family.[18]

In Japan, when children attend elementary school, they generally commute to and from school by themselves, without their parents. Moreover, when children reach the upper grades of elementary school, they begin commuting alone to attend cram school or after-school lessons. Since some finish cram school after dark, at times a child can return home after 10 p.m. This is why commuting to cram school prompts *keitai* ownership for an elementary school child.

In Japan, the perception that public safety is deteriorating[19] has led to the falling age in *keitai* ownership as parents turn to *keitai* as a practical way of ensuring child safety. In addition to crime prevention alarms, GPS (geographic positioning system) functioning is available with *keitai* marketed for children that allows the subscriber (in most instances the parent) to access the location information of a particular *keitai*.

If we reflect on this, however, it is clear that a child's safety cannot be secured by their mere possession of a *keitai*. In light of the recent shortage of public telephones, *keitai* may be useful for when a child needs a parent to pick him or her up after cram school.[20] However, in reality, carrying a *keitai* does not decrease the chance of a child being victimized. Even if a child's whereabouts are known at all times through the use of such *keitai* features as the GPS tracker, this is not for a child's safety, but is purely for the sake of parental peace of mind.

Let us return to the opening quotation for this part in which a ten-year-old girl is persuading her mother to buy her a *keitai*. The child is cognizant of the "meaning" of the *keitai*.[21] In other words, she understands that it is not the child's safety but the parent's sense of the child's safety that encourages the trend towards *keitai* marketed for children. So the question we are left to ask is Why is a parent's fears about the child's safety diminished through the acquisition of a *keitai*?

In response to the popular belief that discipline in the home has deteriorated, Teruyuki Hirota—who has researched material from the Meiji period up until present—counters with a theory from a perspective of a social history. He argues that parents today are in fact more passionate about their children's education and discipline. At the same time, as a result of the strong emergence of "family centrism," he claims that social standards have reached the point where "the family as a unit must wholeheartedly accept ultimate responsibility for their child's education."[22]

Because of this strong concern for the child,[23] parents habitually feel the necessity to make arrangements for the child's daily activities, and more-over, using the technology/*keitai* that provides them this kind of control, they come to adopt the intended uses of its various functions. In other words, the reason for the popularity of *keitai* marketed for children is this "surveillance" of a "family who educates," under the pretext of "the dete-rioration of public safety."[24]

Here, the dual meaning of "surveillance"—both "care" and "control"—as emphasized by David Lyon is significant.[25] The act of secur-ing a child's safety via *keitai* is gentle supervision performed "with concern for the sake of another" or, in other words, "caretaking."[26] It is difficult to reject "surveillance" per se, since it is an expression of "care." How-ever, the party whose actions are under "surveillance"—as "control"—is established from the start: the parent confirms a child's location—never the reverse. Moreover, electronic care/control via *keitai* doesn't afford the mutuality of a copresent situation—the child is not able to read the atmo-sphere of a parent's "surveillance." Furthermore, the act of confirming whether a child has made it safely to cram school can easily be associated with checking whether a child is cutting cram school—or the "control" of a child's actions.[27]

Even when they are apart from their children, parents either habitu-ally exercise "care/control" over them or they have no choice but to do so. For this reason, they encourage their children to own a *keitai* that has no direct effect in ensuring their "safety." Due to the use of *keitai*, many people, for some time now, feel that they are working around the clock. Likewise, the *keitai* is depriving parents of "breaks" from parenthood.

MOM IN THE POCKET

> *Author:* Is there any particular instance when you were glad you had your child carry a *keitai*?
>
> *Mother of nine-year-old daughter:* Well, my daughter called me while she was out crying, "I lost my bicycle key!" So I told her, "I'll be there right away, so wait right there." I was glad she was carrying a *keitai* then.

In their use of the term *remote mothering*, Lana Rakow and Vija Navarro shed light on the fact that by carrying a cellular phone, a mother can/must carry out her duties of motherhood at any time and any place.[28] The same tendency is seen in Japan, where the use of the *keitai* has become an oppor-tunity to reinforce the gender order.[29] Now, what happens when a child comes to possess a *keitai*?

Firstly, "remote mothering" would be further enhanced for the mother. This is because the opportunities for contact would increase with both parties in possession of a *keitai*. When children have *keitai* many fear that they will access *deai-kei* (encountering/dating) sites, pornography, and violence or that they will run into problems with "dangerous others." *Keitai* have also drawn concern from parents due to their use as a tool for bullying among children. The most common approach to dealing with these problems is for parents to restrict or regulate use.

I do believe that it is important for parents to take an interest in their child's *keitai* use, provide guidance, and develop shared standards. The fact, however, that the parents have made the choice to purchase a *keitai* for their child and the fact that they control their child's use of their *keitai* does serve to expand the sphere of care/control for "families who educate." Moreover, because one characteristic of *keitai* is that they facilitate connection directly with an individual, given the current state of affairs, aren't they likely to strengthen burdens imposed on a mother? In actuality, parent-child communications via *keitai* are predominantly made through the traditionally stronger communication channel of mother and child, not father and child.[30]

The dialogue at the beginning of each section was obtained from interviews conducted in November 2006 of mothers of elementary or junior high school students who either made their child carry a *keitai* or are considering getting a *keitai* for their child. The most interesting statements from the interviews came from mothers who have their children carry *keitai,* who said, "They call me with much greater frequency than I expected." The mothers stated that their children would call to ask questions such as, "Where are the snacks?" or "It's starting to rain, so should I take in the laundry?" Of course, there are instances—as in those presented in the opening of this section—where the "utility" of the *keitai* can be confirmed. Yet what would a child have done in such circumstances in the pre-*keitai* era?

The *keitai* has increased the amount of "nonurgent business" (matters that warrant a phone call, but that are not of any urgency).[31] Phone calls are primarily motivated by some concrete matter, yet the bar for what counts as "urgent business" has lowered, so even what was formerly seen as "a matter unworthy of a phone call" has now become "business." Given this tendency in *keitai* usage, it comes as absolutely no surprise that a child uses a *keitai* to call his or her mother to make inquires on the smallest issue.

Giving their opinions about "nonurgent business" calls from their children while they are at work, mothers responded that "It's okay because I understand their circumstances" or "It can't be helped," while feeling that the calls are slightly "bothersome."

If the *keitai* serves as a "remote control" for a mother, for a child, the *keitai* is a "hotline to mom." Children spend each day with their dependable moms tucked in their pockets. If any trouble arises, without having to exercise their own judgment, children can just call the "hotline to mom." Their moms are always in their pockets.

CONCLUSION

Upon observing the transformation of Japanese families, I have considered the role of the *keitai* in familial and parent-child relationships. Through the material presented previously, I believe we can grasp one aspect of how embedded the *keitai* is in the specific cultural context of Japanese families.

When mobile media first became popular, the blurred distinction between public and private domains was the issue of the day. Private conversations that spill into public spaces, business persons on vacation making work-related calls, personal calls taken at the workplace, calls from work or friends that disrupt family togetherness were topics of concern. In present-day society in which it has become a given that one can be contacted at any time and place by a desired party—in other words "intimate others"—through mobile media, the "I" in the equation cannot be disconnected from "intimate others."

In my previous work, I defined *keitai* addiction among the younger generation as "a kind of a rite of passage."[32] The use of *keitai* e-mail for close to two hours a day, the daily exchange of several hundred *keitai* e-mails, the anxiety if one doesn't have a *keitai* nearby—these are all activities that lead to parental anxiety. The reason why young people become addicted to using *keitai* is that through their *keitai*, they can constantly get information and feedback about themselves—what their friends think about them; how they should behave in specific situations according to the status they hold in their social group, and so on.[33] To confirm "who I am," young people use mobile e-mails a hundred times a day.

Thus, as one "becomes an adult," one will go on to "graduate" from his or her *keitai* addiction. In actuality, the highest users of *keitai* e-mail are high school students, and the majority of those who make private calls are in their early twenties. While the latter may never part with their *keitai,* most will be released from their addictions. So how then do children who hold their "mom in the pocket" go on to "graduate"? Society associates certain problems with our current *keitai*-saturated era, denouncing the *keitai*-dependent parent and child and the *keitai*-dependent family: "Today's children are overprotected" and "It's also important that parents become independent of their children." However, it is important to objectively assess how distinct the child's "process of independence" and parent-child relationships really are compared to the pre-*keitai* era.

At the same time, this should lead to a reflection upon ourselves, who do not hope to be disconnected from "intimate others"—or conversely, ourselves, who do not seek to have connections to "other people" with whom we are not intimate. This in turn leads to broader questions about our society. Of course, it is not mobile media alone that have given us such preferences and orientations. Rather, the popularity of mobile media is expressly due to its effectiveness in manifesting these preferences and orientations, and is a measure of them.

The situation in Japan—where the rate of diffusion of *keitai* for early teens exceeds 50 percent and possession among elementary school children continues to rapidly expand—is extremely unique among countries or regions where the diffusion of the cellular phone arose as the first voice-based personal communication medium amidst the lack of advancement of landline infrastructure. Moreover, families who are "connected by *keitai*" must appear strange from the point of view of societies where cellular phones are not needed as a means of communication either because of the strong concern that the electromagnetic radiation will impact children's health or because children are rarely unattended by a guardian outside the home.

Even concerning the most intimate relationships between family members or parents and their children, the management or, in other words, the appearance that a family must manage via the new media form of *keitai* is in a sense extremely unique to Japan. Yet even other regions or countries where "family crises" are perceived as problematic, and, moreover, everyone who is part of contemporary society in which connection with "intimate others" at any time of the day is sought should be able also to share such familial "intent" and "effort." What we can see by examining the case of the *keitai* in the specific context of Japanese society is the very structure of contemporary society embedded with mobile media.

NOTES

1. *Keitai* is a shared nickname for mobile phones in Japan. Hereafter, when discussing the cellular phone in the context of Japan, the term *keitai* will be used. See Misa Matsuda, "Discourses of *Keitai* in Japan," in *Personal, Portable, Pedestrian*, ed. Mizuko Ito, Daisuke Okabe, and Misa Matsuda (Cambridge: MIT Press, 2005), 19–39.
2. Mizuko Ito, "Introduction: Personal, Portable, Pedestrian", in *Personal, Portable, Pedestrian*, ed. Mizuko Ito, Daisuke Okabe, and Misa Matsuda (Cambridge: MIT Press, 2005), 1–16. See also Tetsuo Kogawa, "Beyond Electronic Individualism," *Canadian Journal of Political and Social Theory*, 8(3) (1984): 15–20.
3. Tomoyuki Okada, "Youth Culture and the Shaping of Japanese Mobile Media," in *Personal, Portable, Pedestrian*, ed. Mizuko Ito, Daisuke Okabe, and Misa Matsuda (Cambridge: MIT Press, 2005), 41–60; Howard Rheingold, *Smart Mobs: The Next Social Revolution* (New York: Perseus, 2002); Jon Agar, *Constant Touch: A Global History of the Mobile Phone* (London: Icon Books, 2003).
4. Matsuda, "Discourses of *Keitai*."
5. Misa Matsuda, "Mobile Shakai no Yukue (The Future of Mobile Society)," in *Keitai-Gaku Nyumon*, ed. Tomoyuki Okada and Misa Matsuda (Tokyo: Yuhikaku, 2002), 205–27.
6. Rich Ling, *The Mobile Connection: The Cell Phone's Impact on Society* (San Mateo, CA: Morgan Kaufmann, 2004); Timo Kopomaa, *The City in Your Pocket: The Birth of the Mobile Information Society* (Helsinki: Gaudeamus, 2000).

7. The quotes at the beginning each section are from the interviews of mothers with school-aged children in Tokyo in November 2006. For more details about the interview results, see Misa Matsuda, "Children with Keitai: When Mobile Phones Change from 'Unnecessary' to 'Necessary,' " *East Asian Science, Technology and Society*, 1(4) (forthcoming).

8. Misa Matsuda, "Wakamono no Yujin Kankei to Keitaidenwa Riyo (Friendship of Young People and their Use of Mobile Phones)," *Shakai Johogaku Kenkyu*, 4 (2000): 111–22; Misa Matsuda, "Mobile Communication and Selective Sociality," in *Personal, Portable, Pedestrian*, ed. Mizuko Ito, Daisuke Okabe, and Misa Matsuda (Cambridge: MIT Press, 2005b), 173–94. See also Leopoldina Fortunati, "Italy: Stereotypes, True and False," in *Perpetual Contact: Mobile Communication, Private Talk, Public Performance*, ed. James E. Katz and Mark Aakhus (Cambridge: Cambridge University Press, 2002), 42–62 (51).

9. Yoriko Meguro, *Kojinka suru Kazoku (Individualization of the Family)* (Tokyo: Keiso Shobo, 1987).

10. Also see Brian McVeigh, *Nationalism of Japan: Managing and Mystifying Identity* (New York: Rowman and Littlefield, 2004).

11. Ulrich Beck and Elisabeth Beck-Gernsheim, *The Normal Chaos of Love* (London: Polity Press, 1995); Anthony Giddens, *The Transformation of Intimacy: Sexuality, Love and Eroticism in Modern Societies* (London: Polity Press, 1992); Stephanie Coontz, *The Way We Really Are: Coming to Terms with America's Changing Families* (New York: Basic Books, 1997).

12. Ulrich Beck and Elisabeth Beck-Gernsheim, *Individualization: Institutionalized Individualism and Its Social and Political Consequences* (London: Sage, 2001).

13. Yet the ratio of unmarried is rising; in a survey taken between 1982 and 2005 of unmarried Japanese adults of ages 18 to 34, the percentage of those who answered that they do not intend to get married was less than 10 percent. See National Institute of Population and Social Security Research, "The 13th Basic Survey of Birth Trends" (2006).

14. Yoriko Meguro (1987), v.

15. Regarding the increase in excessive dialogue in the Japanese family, see Hidetoshi Kato, *Kurashi no Sesoshi (Social History of Everyday Life)* (Tokyo: Chuko Shinsho, 2002); Matsuda, "Mobile Communication and Selective Sociality."

16. See NHK Broadcasting Culture Research Institute, *Nihonjin no Seikatsujikan 2005 (Time Use of Japanese in 2005)* (Tokyo: Nihon Hoso Shuppan Kyokai, 2006).

17. Mobile Society Research Institute, *Mobile Shakai Hakusho (White Paper on Mobile Society)* (Tokyo: NTT Shuppan, 2005).

18. See Matsuda, "Children with Keitai." For communications between high school students and their parents via *keitai*, see Daisuke Tsuji, "Wakamono niokeru Idoutaitsushin Media no Riyo to Kazokukankei no Henyo (Youth Use of Mobile Media and Changes in Family Relationships)," *Kenkyuo Sosho: Institute of Economic and Political studies in Kansai University*, 133 (2003): 73–92.

19. Since the latter half of 1990s, "the deterioration of public safety" has been drawing concern in Japan. This "deterioration," however, is not necessarily an objective fact but it reveals a certain state of moral panic. See Kouichi Hamai, "Nihon no Chian-akka-Shinwa ha Ikani Tsukuraretaka (How 'the Myth of a Collapsing Safe Society' Has Been Created in Japan)," *Hanzai Shakaigaku Kenkyu*, 29 (2005): 10–26.

20. In 2006, the number of public telephone facilities was 393,000—less than half that of 1991, when the number peaked at 833,000. See Ministry of Internal Affairs and Communications (2006).

21. Although it is not part of this discussion, another "meaning" of *keitai* marketed for children is without doubt consumerism. For instance, the child-oriented *keitai Papipo* that was created through toy manufacturer Bandai offers an abundance of contents desired by children, such as games and fortune-telling on a handset adorned with popular characters such as *Tamagochi* and Hello Kitty. This "cute" *keitai* that was planned and developed with a target ages of seven to twelve is to activate the saturated *keitai* market—and children's "safety" is merely a guise.

22. Teruyuki Hirota, *Nihonjin no Shitsuke ha Suitai Shitaka (Did the Discipline of Children Decline among Japanese?)* (Tokyo: Kodansha Gendai Shinsho, 1999), 127.

23. Naturally, it is assumed that due to the decrease in the rate of births and infant mortality, there is a general sense of value that "we must bring up our precious few children even better."

24. Concerning mobile phones and surveillance, see Nicola Green, "Who's Watching Whom?" in *Wireless World: Social and International Aspects of the Mobile Age*, ed. Barry Brown, Nicola Green and Richard Harper (London: Springer, 2001), 32–45.

25. David Lyon, *Surveillance Society: Monitoring Everyday Life* (Buckingham, UK: Open University Press, 2001).

26. Families who make their senior members carry a *keitai* can be viewed from a similar perspective. See Matsuda, "Mobile Shakai no Yukue (The Future of Mobile Society)."

27. Misa Matsuda, "Camera tsuki Keitai to Kanshishakai (Keitai with Digital Camera and the Surveillance Society)," *Biomechanisms Gakkai-shi*, 28(3) (2004): 129–35.

28. Lana R. Rakow and Vija Navarro, "Remote Mothering and the Parallel Shift," *Critical Studies in Mass Communication*, 10 (1993): 114–57.

29. Matsuda, "Mobile Communication and Selective Sociability"; Shingo Dobashi, "The Gendered Use of Keitai in Domestic Contexts," in *Personal, Portable, Pedestrian*, ed. Mizuko Ito, Daisuke Okabe, and Misa Matsuda (Cambridge: MIT Press, 2005), 219–36.

30. Tsuji, "Wakamono niokeru Idoutaitsushin Media no Riyo to Kazokukankei no Henyo (Youth Use of Mobile Media and Changes in Family Relationships)."

31. Misa Matsuda, "Ido-denwa Riyo no Case Study (A Case Study on Mobile Telephone Usage)," *Tokyo Daigaku Shaka Joho Kenkyusho ChosaKenkyu Kiyo*, 7 (1996): 167–89.

32. Misa Matsuda, "Keitai (Mobile Phone) and Youth in Japan," *International Symposium: The Social Use of Anthropology in the Contemporary World*, National Museum of Ethnology, Osaka, October 28, 2004.

33. Ichiyo Habuchi, "Keitai ni utsuru 'Watashi' (The 'Me' Reflected in the Keitai)," in *Keitai-Gaku Nyumon*, ed. Tomoyuki Okada and Misa Matsuda (Tokyo: Yuhikaku, 2002), 101–21.

7 Purikura as a Social Management Tool

Daisuke Okabe, Mizuko Ito, Aico Shimizu, and Jan Chipchase

Since the late nineties, Print Club, or *purikura*, sticker photos taken at photo booths have been widely established among Japanese female high school students as a means of visual archiving and communication. According to Okada, the first *purikura* booth appeared in July 1995, and by October 1997 there were 45,000 units nationwide.[1] *Purikura* stickers are taken and modified with graffiti in *purikura* booths, then printed out on the spot. The final *purikura* sticker sheet is cut into separate stickers, which are then stored in a special notebook for this purpose, called the *puri-cho*. Leftover stickers are stored in a container, a *puri-kan*, for trading with friends at a later time. If rare or highly valued, stickers are placed on mobile phones, pencil cases, and pocket mirrors.

In recent years, it has also become common to send *purikura* images to mobile phones for archiving purposes. *Purikura* created a new modality for the archiving and sharing of visual information, as well as new means of communication via image sharing.[2] *Purikura* are popular among teenage girls, particularly during the high school years. The majority of Japanese teenage girls have used *purikura* booths at least once. A November 2005 survey of girls aged ten to eighteen found that *purikura* was one of the most popular media activities among this demographic.[3] With 43.6 percent of respondents choosing *purikura* when asked to select their favorite activities (multiple answers allowed), *purikura* was more popular than reading comics (a well-established pastime among Japanese youth), which garnered votes from 40.1 percent of respondents.

Purikura are part of an ensemble of different mobile and location-based visual archiving and sharing practices that young people in Japan mobilize in their everyday lives. They are a relatively recent addition to this media and communication ecology, and are part of a much longer trajectory in the evolution of visual culture. Marita Sturken and Lisa Cartwright, in their analysis of visual communication, suggest that we are living in a visual culture in which we constantly communicate and influence each other by viewing images mediated by image technologies.[4] The popular cultures of youth in Japan are saturated with visual and media culture; this visual culture is not only a source of media and entertainment, but is an integral part of the stories that they tell about their everyday lives to themselves and others. For decades now, the camera has been a common mobile media device for capturing personal visual media. The advent of services such as *purikura* and the digital camera has accelerated this trend towards visual culture becoming integrated into everyday chronicling and communication.

The late nineties saw the convergence of the practices and technologies surrounding digital cameras, *purikura,* and mobile phones. The adoption of camera phone technology occurred in tandem with the spread of *purikura.* Like *purikura,* cameraphone photography was adopted as mode of lightweight, everyday photography that made images more pervasive in the lives of young people.[5] In our research on camera phones, we found that they were used to capture much more mundane and fleeting everyday moments, in comparison to traditional amateur photography. Young people would capture photos of food, friends, and their local settings and share them casually with one another, rather than reserving photography for special occasions and long–term visual archives.[6] We see *purikura* and camera phones as occupying different niches in the visual cultures of Japanese youth, but both are part of a broader trend in mobile visual media. Images are becoming increasingly integrated into personal documentation and social sharing as visual capture and sharing devices are more pervasive in people's pockets and diverse environments.

This paper builds on our earlier work on mobile communication and cameraphone use by reporting on an ethnographic study of *purikura* use conducted in the summer of 2006, at a point when *purikura* was a well-established element of teenage girls' visual culture practices. Our goal is to describe everyday practices of *purikura* taking, archiving, and sharing, and to analyze how this form of visual media becomes a mechanism for young people to document and display their personal identities, social status, and networks.

PURIKURA IN CONTEXT

The adoption of *purikura* in Japan is overwhelmingly centered on teenagers, with high school girls being at the core of this new visual and communication idiom. The relation between *purikura* and the social life of high school girls is related to longstanding dynamics of social identity and status construction in the high school years. High school is a period when young people develop their identities and hierarchies of social status in the relatively closed networks of high school peer relationships. The status negotiations among high school students often involve the mobilization of consumer and media culture in order to display identity and status.[7] High school teens are going through a period of their lives in which they have the power to construct an informal social milieu by evaluating each other's social status with a system that is disparate from that of their teachers and parents. Although the specific media, technologies, and cultural markers have changed over time, the underlying dynamics of peer status negotiations in high school have been resilient over time. High schools today, even those in Japan, reflect many of the dynamics that were documented in youth studies in the sixties and seventies.[8] All of these studies of youth culture and status describe the complex social negotiations that young people engage in to display their identities and establish boundaries that define different social groups and statuses. We believe that modern media serve an important function as markers of membership and social status within these groups.

Prior work on mobile and visual media in Japan provides a context for understanding some of the specificities of *purikura*. For example, Matsuda has described how Japanese teens are able to maintain selective friendship networks, where youth are increasingly able to keep in touch with old friends even though they many not share a neighborhood or a school.[9] Communications media such as the mobile phone are central to this dynamic; they provide an infrastructure for keeping existing social networks alive. Our prior work on text messaging among Japanese teens has also described how communication media become a "site" for enacting friendships and romantic relationships, providing both an arena for interaction and an archive of past social exchange. Young people have a sense of inhabiting a shared space of interaction through ongoing text-messaging exchange.[10] Cameraphones enter this stream of exchange as a visual archive of an individual's personal viewpoint and social encounters.[11] In a similar vein, Izumi Tsuji has done ethnographic research on the link between media and friendship. He conducted a sociometric test of members of an idol fan group and analyzed how their peer network was embedded in their mobile phone's memory.[12]

Although there is a growing literature on Japanese youth and new media, there is very little existing scholarly work on *purikura*. Nobusyohi Kurita did some early work on the social functions of *purikura*. He observed how young people take *purikura* when they are engaged in social gatherings out of home and school, and that the photos become a mechanism to commemorate events and affirm friendships with peers.[13] Laura Miller's more recent work on *purikura* looks at the photos as an expressive medium that is embedded in the specific cultural practices and literacy of teenage girls. Drawing from fieldwork on the street cultures of high school girls, she examines particular genres of *purikura* graffiti such as *"hengao"* (funny faces), *"kimo-puri"* (gross *purikura*), and *"yaba-puri"* (dangerous *purikura*). She argues that these forms of *purikura* can be construed as gender parody, an act of rebellion against the social expectation that women should act "cute." She also describes the ways in which the cultural styles of young women have been at the forefront in defining contemporary visual culture, and sees *purikura* as a recent demonstration of girl culture's defining role.[14]

Our work builds on the prior work on Japanese youth and new media by looking specifically at *purikura* as a form of girl- and youth-driven bottom-up innovation in visual culture. We, however, focus specifically on the role of *purikura* in making visible youth social networks and status negotiations. *Purikura* perform multiple roles, including the expressive functions that Miller explores, as well as the more social functions of negotiating status within the peer networks of high school relationships. This paper examines this latter role of *purikura*.

FIELDWORK

Fieldwork for this study included a mix of observations and interviews. In July and August 2006, we observed *purikura* centers and game centers with *purikura* booths for three days at different times of days in urban areas such as Shibuya, Kichijoji, and Yokohama. In addition to observations, we conducted spot interviews with 18 users exiting the booths, asking questions regarding how many people they took *purikura* with, their relationship with each other, the purpose of their *purikura* use, whether they owned a *purikura* album, the content of graffiti on their *purikura* sticker photos, and whether they sent the *purikura* images to their mobile phone. We also interviewed a staff member at each of the three game centers we observed regarding the history of how *purikura* booths were incorporated at their center, the shift in user demographics over time, and changes in booth equipment. We also interviewed a representative of a *purikura* booth manufacturer regarding the social factors involved in *purikura* software design.

In addition to observations and spot interviews at *purikura* locations, we conducted in-depth interviews and shadowed young women, who were frequent *purikura* users. We interviewed sixteen young women: one junior high school student (age fifteen), twelve high school students (aged sixteen to eighteen), and three college students (aged eighteen to twenty-one). The junior high school student lived and went to school in Kanagawa Prefecture; the high school and college students all lived and went to school in Kanagawa and Tokyo Prefectures. The high school students attended schools that were ranked as being of medium level or preparatory. We also interviewed two male college students (aged nineteen and twenty-two) regarding their *purikura* use. The interviews and shadowing occurred between May 2006 and September 2007.

THE SOCIAL USES OF *PURIKURA*

Overview

For teenage girls, *purikura* are a routine activity when they get together with their friends. While the frequency varied, many interviewees stated that they take *purikura* every time they go out with their friends. *Purikura* is a tool to visualize the fact that users hung out together and to commemorate the event. Interviewees also tend to take *purikura* after school-related gatherings such as club activities and culture festivals. While taking *purikura* with immediate and extended family members is rare, it does happen on occasion. The number of *purikura* sticker photos taken varies depending on the individual and context, but it is not uncommon for multiple *puri-kura* sessions to occur over the course of the day. One interviewee explains: "We walk around and duck in for one *purikura* session, then go for another round after lunch." It is also common for two to three consecutive takes to occur when four or more people are involved. This not only increases the number of sticker photos per person but also facilitates cost splitting.

Purikura are rarely, if ever, taken alone. During our observations in the field, we didn't see any women using *purikura* booths by themselves. We only observed one group of three male high school students who came as to take individual photos. They took *purikura* one by one to make sticker photos as a kind of calling card. The *purikura* we saw in *purikura* albums all had two or more participants, supporting our observations that *purikura* are almost always taken with others. Young people take *purikura* with a wide range of friends. When asked whom they take *purikura* with, most informants replied that there were too many to list. Many of the photos in *purikura* albums were taken with the same group of friends from school, but there were many others taken with people belonging to a wide range of groups, such as other classmates, friends from clubs and activities, students from different grades at school, and relatives.

Taking photos is just one part of the overall ecology of *purikura*-related activity. After the photos are taken, young women will cut them up and array them in elaborate *pur-cho*, which are carried around with them on social occasions together with a *puri-kan* containing duplicate *purikura* stickers. All of our informants had *purikura* albums and *puri-kan*. One young woman explains, "Whenever you ask to see someone's *puri-cho*, no one ever fails to produce one [K, female, eleventh grade, age seventeen]." *Puri-cho* are not simple collections of stickers, but are generally painstakingly designed to represent the owner's taste and aesthetic sense (Figure 7.1). These albums function not only as a personal diary and archive, but also as a social medium to be shared among friends.

Figure 7.1 Left, *puri-cho* of a twelfth grader; right, puri-cho eleventh grader. Source: Author.

Purikura activities differentiate markedly along gender lines. *Purikura* users are overwhelmingly female, and boys generally only take *purikura* if they are with a girl or group of girls. Many interviewees indicated that it was rare for them to take *purikura* with boys: "I don't take *purikura* with guys. Even if I do, it's usually on the spur of the moment, like after graduation ceremonies [K, female, eleventh grade]." In our observations at *purikura* locations, it was most common to see groups of girls, but we also observed many couples taking *purikura* together. The two male college students we interviewed had several *purikura* sticker photos in their planners and stuck on their mobile phones. However, these photos were either taken with their girlfriends or were group photos taken while out drinking with their friends. Of all the *purikura* we observed during interviews, we saw only one taken of a group of boys. The interviewee explains that it had been taken "as a joke when friends from high school came to visit [S, male, sophomore]." Unlike girls who take *purikura* more frequently, and carry them around in albums and *puri-kan*, for boys, it is a much more peripheral activity, and one that they engage in largely to associate with girls.

Although *purikura* are pervasive among teenage girls, there are certain subcultures more closely identified with *purikura* activity. When asked to characterize people who frequently take *purikura*, informants tended to list teen girls who are considered "gals." Gals or "*kôgyaru*" (little gals) is a term that came into currency in the late nineties to describe street savvy teenage girls who sported loud fashions and a bad-girl demeanor. *Purikura* culture became emblematic of gal street culture,[15] and continues to be associated with it.

> Gals are pretty into *purikura*. When we go take *purikura*, we only take about three at most, but gals take six to seven. [Pointing to sticker photo in *purikura* album] This girl takes eight to ten in one go. [H, female, twelfth grade, age eighteen]

Purikura use is highest among junior high school and high school students, especially among the latter. Young people who have more pocket money and freedom to take outings away from parents and school activities are the most frequent *purikura* users. The frequency of *purikura* engagement also varies depending on the kind of social networks that young people occupy. High school students who are in many clubs and stay in touch with childhood friends take *purikura* with higher frequency. Conversely, when young people start to participate in social networks removed from high school, and socialize less with their classmates, *purikura* usage drops. The following interviewee indicates this trend in her life.

> Actually, I don't take *purikura* very often now. I started being active in music circles around May or June, and they're mostly guys in their 30s and 40s. Before, I wasn't busy with music so I would hang out with friends. Well, I think I took *purikura* every time I hung out with friends. [Y, female, twelfth grade, age eighteen]

It follows, then, that the frequency of *purikura* use and *puri-cho* updates differs greatly between high school and college students. Transcripts of interviews with college students indicate that the frequency of their *purikura* use drops dramatically after graduation from high school, and *puri-cho* are rarely updated in college. When college students take *purikura*, the majority sandwich the photos between the pages of their planners or merely archive them on their mobile phones; it is extremely rare for them to maintain *puri-cho*.

Figure 7.2 is a *puri-cho* of a college freshman. The interview was conducted in June, only two months after the interviewee began attending college. She had no entries after her high school graduation ceremony on March 25. In addition, the graduation ceremony entry has no textual annotation. This example is a good illustration of the *purikura* album's particular function as a record of the owner's everyday activities as well as a communication tool with others. Upon entering college, the structure of peer networks changes. Rather than social relationships revolving around classes and clubs shifts to more individualized mode, college students are charting their daily activities with more independence. The commutes home after class, weekend activities, and leisure activities also undergo a drastic transformation during this time, leading to fewer opportunities for activities that had been contexts for *purikura* use and less unstructured time between classes when classmates hang out together. College students still take *purikura* once or twice every two to three months, when they make a plan to get together with their friends. This is significantly less than the high school students' frequency of two to three times every month. College students generally either place the photos between the pages of their planners or paste them on available space on their planners (Figure 7.3) and will not carry around a dedicated *puri-cho*.

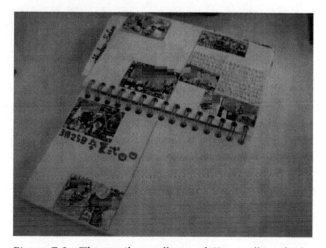

Figure 7.2 The *purikura* album of K, a college freshman. Entries cease after March 25, when K was still a high school senior. Source: Author.

Figure 7.3 The planner of K, a college freshman. *Purikura* are placed on the lower margin of pages in her scheduler. Source: Author.

Having provided an overview of how *purikura* operate in the lives of Japanese youth, we turn now to an analysis of three dimensions of the social practices surrounding *purikura*: 1. Sharing of everyday social moments, 2. Self-expression, and 3. Peer networking.

Sharing Everyday Social Memories

> *Interviewee:* We took this one when we went to show our support at a soccer match. We always make sure to take *purikura* when we're dressed funny. It was right before we took the bus to the stadium—where was it? I think it was in Shibuya. We wanted to take *purikura* before going to Saitama Stadium, so we met up beforehand in Shibuya.
>
> *Interviewer:* Why? Because you were going to a soccer match?
>
> *Interviewee:* We were wearing uniforms, and that was a big deal.
>
> [K, female, college freshman, age eighteen]

Purikura have taken hold in Japanese youth culture because they occupy a unique role that differs in important ways from other forms of photography and photo archiving. *Purikura* need to be understood based on their primary role commemorating everyday social gatherings. Unlike digital cameras and camera phones, *purikura* booths are attached to particular locations and are optimized to photograph small groups of individuals. *Purikura* are designed so all members of a group can be photographed, in a photo booth with good lighting and interesting backdrops.

Unlike a personal camera, which is directed at others and the external environment, the *purikura* booth camera is directed at the self, photographed together with friends. The *purikura* is in the genre of photography similar to the "group photo" where someone might trot out a tripod and put the camera on a timer. The difference is that *purikura* are taken much more casually, to commemorate routine and everyday moments of social gathering. Young people do not go out to specifically take *purikura* photos. Rather *purikura* are taken before and after activities such as "going to a soccer match," "going to watch a movie with friends," "going to take exams at school," or "going to hang out with a friend for the first time."

The other important characteristic of *purikura* is that users can write graffiti on captured *purikura* images on the spot.[16] After taking the photos, users go to a "graffiti corner" with a touch screen and pens with elaborate features to add backdrops, write on the photos, or decorate with colorful stamps or sparkles. The collaborative annotation of photos is also part of the commemoration process. At *purikura* centers you will see groups of young women laughing and chatting as they annotate their photos, or couples leaning up against each other behind the curtain of the graffiti corner, annotating their "love-puri." The annotation becomes a way of producing shared meaning about the occasion. For example, one group who had come to take *purikura* after exams wrote "All done!" on their sticker photos. Another example is "*hatsu-puri* (first *purikura*)" occasions, where users annotate the sticker photos with graffiti such as "Friends!" to denote the fact that they have now established a friendship. This annotation of "friends" or "*hatsu-puri*" is used specifically to indicate the first time someone went out with particular friend or group of friends. The *hatsu-puri* photo, archived on a particular date in a *puri-cho*, becomes a way of commemorating the memory of a new friendship being formed.

In our spot interviews, eight out of eighteen respondents had sent the *purikura* images they had just taken to their mobile phones, with sixteen having sent *purikura* images to their mobile phones at least once. Although this ability to send the digital photo to a mobile phone is a relatively new one, it is already becoming an established feature of *purikura* culture (Figure 7.4). Although these digital versions are not displacing the *puri-cho*, they are one more way that young people can store and share their *purikura* photos. One informant carried two mobile phones with her when meeting with friends. Her new mobile phone was not compatible with the memory card of her old mobile phone and she wanted to be able to share the *puri-kura* images on her old mobile phone with her friends.

Whether in a mobile phone or in a *puri-cho*, *purikura* function as visual displays of social relations and fond memories of being with friends. All of our interviewees were able to answer questions about the day a particular *purikura* image had been taken without hesitation. The small, portable format of the *purikura*, the *puri-cho*, and the mobile phone *purikura* images all point to a more ubiquitous and low-profile presence of imagery

Figure 7.4   Purikura image archived on a mobile phone
[N, female, twelfth grade, age eighteen]. Source: Author.

in the everyday lives of youth. As we have discussed in our work on mobile
communications, portable media and communication technologies enable
young people to inhabit a shared social space with intimate others, even
when they are not together physically.[17] *Purikura* are part of this general
phenomenon; young people's social memories are always close at hand in
an intimate visual form.

The modification and archiving of *purikura* also serve to layer social meaning and memory onto the photos. Expressing aspects of their lives—what they
saw, what they felt, and whom they spent time with—means that they are
choosing what they want to commemorate and archive for themselves. At the
same time, they are choosing aspects of themselves to show to others, since
albums are regularly viewed among friends. In other words, *puri-cho* are also
tools to control the impressions album owners impart to others and to express
the owners' identity. In addition to graffiti at the photo booth, users add additional modification and annotation when they stick their photos in their *puri-
cho*. Young women include other materials to create a colorful record of their
daily lives, a bricolage of *purikura* combined with other media such as magazine cutouts, movie tickets, and snippets from fliers.

> I glued on stuff like the flier for the movie we saw that day, and pasted
> the *purikura* we took. Other people see this, so I try to make it interest-
> ing and original.
> [I, female, twelfth grade, age seventeen]

> I glue magazine cutouts on my *purikura* album. I think people like that.
> They all say 'Cute!' It's just not good enough to put [only] *purikura*

sticker photos on albums. It's a pretty competitive culture. It's fun when you're doing it though.
[Y, female, eleventh grade, age sixteen]

Puri-cho include photos that the owner of the album has taken, as well as photos that they have received from friends. The albums are thus bricolage in multiple senses—a bricolage of different representations of social relationships attached to self as well as others, and a bricolage of different materials that illustrates and commemorates social gatherings. *Purikura* images and albums become an individual's "expanded self"[18] that circulates through personal social connections. This process is analogous to what we see on social network sites on the Internet, where online space becomes an arena to display identity and "friend-of-a-friend" social connections. *Purikura* is a visual medium grounded in a broadly networked practice of social gathering, sharing, and exchange. This networked commemorative practice is what makes *purikura* occupy a unique role in the visual cultures of Japanese youth.

Self-Expression

Since *purikura* are visible displays of friendships, relationships, and social connections, they become mobilized in the ongoing popularity and group negotiations of youth. Just like fashion, profiles on social networks sites, or musical tastes, *purikura* are expressions of personal identity and relationships as well as mechanisms to negotiate social status and distinction.

The other day, I was shown a *purikura* that was taken in Odaiba with a—how to say it—a girl in our class who's considered to be something of an idol. *Purikura* taken with her is considered 'rare-puri' and an attention grabber.
[S, female, eleventh grade, age sixteen]

We take *purikura* a lot on our way back from club practice, but it's boring if we're always wearing the same thing, so we try hard to mix it up by wearing neckties and ribbons on our uniforms. We usually tie our sweaters around our waists, but we sometimes wear them when we take *purikura*.
[K, female, eleventh grade, age sixteen]

Achieving social distinction is a goal that *purikura* users aspire to from the moment of image capture. For example, they may intentionally make funny faces (*hengao*), discuss what poses to strike in advance, or take *purikura* with "notable" individuals. These activities are motivated by the assumption that the images will eventually be shared and viewed with others. Unique or notable *purikura* images will likely circulate among friends, so girls strategize to make their *purikura* images distinct from the more common, generic ones. For example, we took *purikura* with some female high school students after an interview. These photos were considered "rare-puri" because high school girls seldom take *purikura* with older people and people outside of their friend networks.

Puri-cho design is another way of demonstrating taste and visual connoisseurship.

> On this day, we went to see Harry Potter, but this was a lot of trouble. I clipped [the cover images of] the back issues of *Screen* [a movie guide magazine] one by one and pasted the cutouts. The images of back issues were the same size as *purikura* sticker photos, so I thought they were perfect. When I show this to my friends, they say they would have never thought of that layout. Getting comments like that make me feel a bit superior. I feel proud because I'm probably the only person doing this kind of thing. It's fun to do things other people don't do.
>
> [Y, female, eleventh grade, age sixteen] (See Figure 7.5)

This interviewee indicates how female high school students make an effort to have uniquely designed *purikura* albums. Deciding what kind of *purikura* image to take and how to design the *puri-cho* are important dimensions of the social practices surrounding *purikura*. These practices are forms or personal identity expression that also help people achieve a certain status among their peers. Creating a unique *purikura* album that displays creativity and taste in the context of social connectedness translates to superior status in a peer networks.

Figure 7.5   Designed *purikura* album with images of magazine back issues juxtaposed with *purikura* sticker photos (Y, twelfth grade). Source: Author.

Interviewees described how they were interested in how people outside of their own peer networks were creating and displaying *purikura*. They were curious about the *puri-cho* of upperclassmen in the same club or social groups outside of school. Gaining access to new *purikura* capture, annotation, and archiving techniques from other social groups can lead to status in one's own local social network.

> Sometimes, I show my *purikura* album to upperclassmen and under-classmen. Upperclassmen have pretty interesting albums, so I like seeing theirs. I only hang out with them at student council meetings. There aren't very many students in the council, so I haven't seen a lot of al-bums, but I've thought of studying the kind of poses they make. I think they do things a bit differently from my classmates. The upperclassmen make poses that are cool, or really just unique. They're fun. So I sug-gest making similar poses when I'm taking *purikura* with classmates. That's how it spreads.
>
> [Y, Female, tenth grade, age sixteen]

Being in the know about a new technique or style confers social status; these new approaches to styles of graffiti or *hengao* spread quickly among different social groups if they come from higher status groups. The new techniques lose their position as status symbols when they spread widely, and older and more connected youth quickly move onto new innovations in technique and expression. In this way, styles of visual expression in *purikura* share the same dynamics as fashion and other forms of popular cultural engagement in being tightly integrated in the ongoing status nego-tiations of youth.

Peer Networks

> *Interviewer:* How well do you have to know each other to start show-ing *purikura* albums?
> *Interviewee:* Well, you don't need to be that close. Even new classmates say, "Can I have some [*purikura*]?" It's a chance to make friends.
> *Interviewer:* Do you make note of who are friends when you're looking through *purikura* albums?
> *Interviewee:* Oh, yeah. I notice things like, "Oh, she's in a lot of *puri-kura*." Like, "Oh, so-and-so shows up a lot, but so-and-so shows up every so often, too."
>
> [K, female, eleventh grade, age sixteen]

When designing their *purikura* albums, young women are selective about which photos they will display for aesthetic reasons, as well as for social rea-sons. *Puri-cho* are social networks displays, and must be constructed with care. Conversations with peers while perusing these *purikura* albums often center on the subject of peer networks. One interviewee explains, "When we see our friends' *purikura* albums, we talk about things like 'How do you know her?' and 'Oh you hung out with her?' " [K, female, eleventh grade, age seventeen]. *Purikura* function as way of observing a friend's social network, as well as a way of affirming a particular social group's ongoing connections. For example, one interviewee describes how *purikura* help maintain a network of old friendships from junior high school (*genchuu*): "Even after going on to high school, I still meet up with *genchuu* friends four to six times a year, and we always take *purikura* together. *Purikura* plays a big role [in keeping in touch]" [Y, female, twelfth grade, age eighteen].

Again, these status dynamics are similar to what we see on social network sites. danah boyd analyzes the link between social status and media in her study of MySpace. MySpace users manipulate their friend list in their profile to visualize interpersonal relationships and social identities, as well as to strategically manage their "coolness" factor by managing their "Top 8" friends list.[19] Similarly, while *purikura* and *puri-cho* have a private, self-documentation dimension, this goes hand in hand with a communication aspect that is very public in practice. The albums are collaboratively constructed, networked records of their owners' daily lives. Although they are structured as personal narratives, they are built and interpreted in ongoing conversation and collaboration with peers. Like social networks sites in the lives of U.S. teens, *purikura* have become a social standard among Japanese teenage girls, and are a key element of participation in peer groups and negotiations over popularity and status. While it is certainly possible to maintain friendships that are not represented in *puri-cho,* the expectation is that relationships should be visible in these semipublic archives of *purikura.* Just like friends lists on social network sites or where kids sit at lunch, *purikura* are mechanisms to display membership in social groups.

One exception, however, to this relatively public circulation of *purikura* images are photos of couples. *Purikura* that boyfriends and girlfriends take together are generally excluded from *puri-cho* and are usually kept at home or placed in more private locations.

> *Interviewee:* For example, I don't put *purikura* taken with my boyfriend [on my *puri-cho*]. I don't put *purikura* I can't show to other people. I usually keep them in my desk drawer at home. Some of my friends have a *puri-can* for *purikura* with their boyfriends and another *puri-can* for friends.
>
> *Interviewer:* You don't bring a *puri-can* for *purikura* with your boyfriend?
>
> *Interviewee:* Even if I do, I'd only share with close friends.
>
> *Interviewer:* Does your boyfriend ask you not to put his *purikura* on your album?
>
> *Interviewee:* I sometimes get asked not to for a while. I ask him if it's okay.

> *Interviewer:* So does it mean you're really good friends with someone if she gives you *purikura* taken with her boyfriend?
>
> *Interviewee:* That might be true. We do say things like "Make sure to give it to me first thing when you take one [with your boyfriend]."

[A, female, eleventh grade, age seventeen]

As the previous exchange indicates, *purikura* taken with boyfriends and other boys are handled very differently from those taken with girls only. One informant commented, "I can't take *purikura* with X [an older man] I met at a live house. Other girls might say, 'I'm jealous [that you could take *purikura* with X]', so I can't" [Y, female, twelfth grade, age eighteen]. *Purikura* taken with men can lead to envy and jealousy so must be handled with care. In our interviews, we saw very few photos in *puri-cho* of couples. When girls display such images in their *puri-cho*, they are making their dating relationships public to a broad network, and it is rarely done. Showing a close friend a secret *purikura* of a boyfriend is an indicator of trust and closeness. Even more than traditional photographs or cameraphone photos, *purikura* are taken and archived with the presumption that they will be under scrutiny by others. Further, the physical and permanent nature of the *puri-cho* makes the owner cautious about archiving photos that they may not want to associate with down the line. Boyfriend photos are the most obvious instance of this caution, where they do not want to have a past relationship in their "permanent record." These examples of the choices young people make in how they array their social relations in their *puri-cho* are yet another indicator of how deeply embedded the visual culture of *purikura* is in the everyday status negotiations of Japanese teens.

CONCLUSION

Our research has described the role *purikura* play in the ongoing self-expression and social negotiations of Japanese youth. In addition to being a new mode of visual culture and creative play, *purikura* are mechanisms to display taste, social distinction, and status. *Purikura* share many of the social displays of social network sites, the aesthetics of digital media and Japanese girls' culture, and the communication practices of mobile media. As with other forms of mobile media—the planner, the mobile phone, and the personal media player—*purikura* and *puri-cho* are miniaturized, customized, and personal media that keep intimate relations close at hand. Always with you, you are never alone; just as the mobile phone enables persistent ambient contact with friends and loved ones, the *puri-cho* tucked into a schoolbag means that one's social identity is available for easy access and display.

While exhibiting many features that align with recent forms of mobile media, *purikura* occupy a unique niche within the media ecology of youth in Japan. *Purikura* are an accessible, inexpensive form of visual culture that is keyed to the aesthetics and sensibilities of teenage girls. While *puri-kura* in many ways reproduce the existing status and social negotiations that have occupied high school students for decades, they also take visual culture in new directions. Like digital cameras, camera phones, and other new mobile media technologies, *purikura* represent the growing influence and pervasiveness of visual media in the everyday lives of young people in Japan. Further, young women exhibit sophisticated social and media literacies in their production of *purikura* and *puri-cho,* demonstrating the kind of grassroots innovation in visual culture that happens when a media technology is taken up as a new youth medium.

ACKNOWLEDGMENTS

This research was supported by Nokia Research, the Docomo House design cottage at Keio's Shonan Fujisawa Campus, and the Annenberg Center for Communication at the University of Southern California. Elissa Sato translated from Japanese.

NOTES

1. Tomoyuki Okada, "Youth Culture and the Shaping of Japanese Mobile Media: Personalization and the Keitai Internet as Multimedia," in *Personal, Portable, Pedestrian: Mobile Phones in Japanese Life,* ed. Mizuko Ito, Daisuke Okabe, and Misa Matsuda (Cambridge, MA: MIT Press, 2005), 41–60.
2. Nobusyohi Kurita, *Purikura Communication in Mass Communication Study,*55 (1999): 131–52; Laura Miller, "Graffiti Photos: Expressive Art in Japanese Girls' Culture," *Harvard Asia Quarterly,* 7(3) (2003): 31–42; Laura Miller, "Bad Girl Photography," in *Bad Girls of Japan,* ed. Jan Bardsley and Laura Miller (London: Palgrave Macmillan, 2005), 126–41.
3. See http://japan.internet.com/wmnews/20051124/5.html.
4. Marita Sturken and Lisa Cartwright, *Practices of Looking: An Introduction to Visual Culture* (Oxford: Oxford University Press, 2001).
5. Fumitoshi Kato, Daisuke Okabe, Mizuko Ito, and Ryuhei Uemoto, "Uses and Possibilities of the Keitai Camera," in *Personal, Portable, Pedestrian: Mobile Phones in Japanese Life,* ed. Mizuko Ito, Daisuke Okabe, and Misa Matsuda (Cambridge, MA: MIT Press, 2005), 301–10; Ilpo Koskinen, "Pervasive Image Capture and Sharing: Methodological Re-marks," in *Pervasive Image Capture and Sharing Workshop, Ubiquitous Computing Conference,* Tokyo, 2005; Tim Kindberg, Maria Spasojevic, Rowanne Fleck, and Abigail Sellen, "How and Why People Use Camera Phones," 2004, http://www.hpl. hp.com/techreports/2004/HPL-2004–216.html; Nancy Van House, Mark Davis, Morgan Ames, Megan Finn, and Vijay Viswana-than, "The Uses of Personal Networked Digital Imaging: An Empirical Study of Cameraphone Photos and Sharing," in *CHI* (Portland, OR: ACM, 2005).
6. Daisuke Okabe and Mizuko Ito, "Everyday Contexts of Camera Phone Use: Steps Toward Techno-Social Ethnographic Frameworks," in *Mobile Communication in Everyday Life: Ethnographic Views, Observations, and Reflections,* ed. Joachim R. Höflich and Maren Hartmann (Berlin: Frank & Timme, 2006), 79–102.

7. Eckert Penerope, "Adolescent Social Categories, Information and Science Learning," in *Toward a Scientific Practice of Science Education*, ed. Marjorie Gardner, James G. Greeno, Frederick Reif, and Alan H. Schoenfeld (Hillsdale, NJ: Lawrence Erlbaum Associates, 1990), 203–17; Paul E. Willis, *Learning to Labor: How Working Class Kids Get Working Class Jobs* (New York: Columbia University Press, 1977); Juliet B. Schor, *The Overspent American: Upscaling, Downshifting, and the New Consumer* (New York: Basic Book, 1998); Murray Milner, *Freaks, Geeks, and Cool Kids: American Teenagers, Schools, and the Culture of Consumption* (New York: Routledge, 2006).

8. Howard Becker, *Outsiders: Studies in the Sociology of Deviance* (Glencoe: Free Press, 1963). Dick Hebdige, *Subculture: The Meaning of Style* (New York: Routledge, 1979).

9. Misa Matsuda, "Mobile Communication and Selective Sociality," in *Personal, Portable, Pedestrian: Mobile Phones in Japanese Life*, ed. Mizuko Ito, Daisuke Okabe, and Misa Matsuda (Cambridge, MA: MIT Press, 2005), 123–42.

10. Mizuko Ito and Daisuke Okabe, "Technosocial Situations: Emergent Structuring of Mobile Email Use," in *Personal, Portable, Pedestrian: Mobile Phones in Japanese Life*, ed. Mizuko Ito, Daisuke Okabe, and Misa Matsuda (Cambridge, MA: MIT Press. 2005), 257–74.

11. Okabe and Ito, "Everyday Contexts of Camera Phone Use: Steps Toward Techno-Social Ethnographic Frameworks."

12. Izumi Tsuji, "Keitai Denwa wo Moto nishita Kakudai Personal Network Chosa no kokoromi: Wakamono no Yujin Kankei wo Chushin ni (Research on Growing Personal Networks on the Memory Bank of a Mobile Phone: From the Viewpoint of Friendship Relations of Contemporary Youth)," *Shakai Jyohogaku Kenkyu* (*Journal of Socio-Information Studies*), 7 (2003): 97–111.

13. Kurita, *Purikura Communication.*

14. Miller, "Graffiti Photos" and "Bad Girl Photography."

15. Miller, "Graffiti Photos."

16. Miller, "Graffiti Photos," and Miller, "Bad Girl Photography."

17. Mizuko Ito and Daisuke Okabe, "Technosocial Situations: Emergent Structurings of Mobile Email Use," in *Personal, Portable, Pedestrian: Mobile Phones in Japanese Life*, ed. Mizuko Ito, Daisuke Okabe, and Misa Matsuda (Cambridge, MA: MIT Press, 2005), 301–10.

18. Schor, *The Overspent American: Upscaling, Downshifting, and the New Consumer.*

19. danah boyd, "Friendster and Publicly Articulated Social Networks," proceedings of *Conference on Human Factors and Computing Systems* (CHI2004, Vienna, Austria; New York: ACM Press, 2004); danah boyd, "Friends, Friendsters, and MySpace Top 8: Writing Community into Being on Social Network Sites," *First Monday*, 11(12) (2006), http://www.firstmonday.org/issues/issue11_12/boyd/.

Part III
Mobiles in the Field of Media

8 Mobile Media on Low-Cost Handsets
The Resiliency of Text Messaging among Small Enterprises in India (and Beyond)

Jonathan Donner

As we reflect on the nature of mobile media, it is tempting to focus on the leading-edge, considering elegant applications in the hands of techno-savvy users, running on advanced devices and over fast data networks in the centers of global digital innovation. There is, however, another equally remarkable side to the rapidly changing environment of mobile communications; thanks to low-cost hardware, prepay tariff plans, and the rapid deployment of network infrastructure in the developing world, hundreds of millions of people who previously could not afford or access landline telephones are purchasing mobile phones. Indeed, the majority of mobile subscriptions are now in the developing world, and, in these markets, lower-cost, simple handsets like the Nokia 1100 are still the leaders by volume.[1] Although these handsets are used predominantly for "traditional" person-to-person voice calls and text messages, there is a variety of mobile media applications which are supported by or designed specifically for even the most basic of mobile telephones. An exploration of these applications raises important questions for this volume, and can add depth and breadth to how researchers and media scholars describe and conceptualize "mobile media" in general.

This chapter begins by describing the limited use of most mobile functions—except for voice calls and SMS/text messages—among small and informal business owners in urban India. It draws on this illustration to suggest that forms of mobile media based on low-cost, ubiquitous SMS features have the potential to be accessible, relevant, and popular among many users in the developing world. Further examples of SMS-based mobile media applications illustrate an important distinction between these systems. While some applications stand alone, others function as bridges to or hybrids of other media forms, particularly the Internet. Over the next few years, these hybrid forms will play an important role in offering flexible, powerful information resources to a sizable proportion of the world's population.

India is one of the world's largest and fastest-growing mobile markets, and the enthusiasm around mobile telephony in India is palpable. In a nation of over one billion, there were 210 million mobile phone subscriptions and only 9.6 million wireline Internet subscribers as of September 2007.[2] Some view these proportions as evidence of a "digital divide," others as both a reflection and reinforcement of a disadvantaged structural position in the global information networks,[3] but mobiles are intertwined with India's current climate of global openness and economic liberalization. They are embraced by some, wanted by many, but not used by all. They are, as Kavoori and Chadha observe, "cultural technologies,"[4] and there is much work to be done, across the span of academic disciplines, to understand how people integrate these iconic devices into everyday life. As such, this brief chapter is not a general review of mobiles in India, nor does it lay out the shape of the industry as a whole.[5]

As of 2007, roughly eight million new Indian users *a month* were buying handsets with a wide range of capabilities. Many urban elites and the new middle class have—or at least desire—higher-end smartphones or feature-rich handsets. According to the Indian telecommunications regulator (The Telecommunications Regulatory Authority of India [TRAI]), there were 49 million mobile Internet subscriptions in late 2007, and there is a vibrant community of developers—multinationals and start-ups alike—developing India-specific content and applications for higher-end phones. Indeed, India has already had its own mobile media scares among users of these higher-end handsets, with multiple Multimedia Messaging Service (MMS) scandals, ranging from doctored, racy images of Bollywood stars to grainy video clips of amorous teenagers, shared virally from mobile phone to mobile phone. But there is a larger group of users, both urban or rural, who by necessity or choice are purchasing secondhand or simple new handsets; perhaps black and white, perhaps lacking General Packet Radio Service (GPRS) (data) functionality. They select prepay plans and seek to keep expenses on ongoing tariffs as low as possible, and well below the average monthly expenditure per subscriber, which in India in 2007 was less than U.S. $7.00 and falling (TRAI). What might mobile media use look like to some of these users?

MOBILE USE BY SMALL AND INFORMAL BUSINESSES IN INDIA

The following discussion focuses on the relatively simple modes of current mobile communication employed by operators of small and informal businesses in India. In many developing countries, the small and informal enterprise sector—microenterprises, the self-employed, small firms with up to five employees—is second only to agriculture as a source of employment and household livelihoods,[6] and many in the development community hope that mobiles (and other ICTs) can help these firms become more productive.[7] As a group, these individuals are neither particularly well off nor without some income; although they cannot afford the feature phones and smartphones favored by urban elites and the new middle class, they are not too poor to afford a mobile.

Data from a 2006 survey about their mobile behaviors suggests that operators of small and informal businesses are one archetype of the low-end handset user. The survey of small and informal enterprises (< 5 employees) was carried out in one of India's largest cities, Hyderabad, and in one of its nearby suburbs, Sangareddi, a town of 50,000 people. The survey's primary focus was ICT use and customer acquisition, the results of which have been presented elsewhere.[8] The survey was not random, nor does it capture rural use cases. However, the survey firm Hansa Research interviewed respondents in numbers proportional to their distribution across five tiers of socioeconomic status, as elaborated by the firm in its nationwide "media readership survey"[9] of Indian households. A total of 317 small businesses with five or fewer employees in Hyderabad (68 percent) and Sangareddi (32 percent) were interviewed, most drawn from various services (55 percent) and from retail endeavors (39 percent). Seventy-eight percent had fixed addresses; the others were roaming (9 percent) or colocated with the home (12 percent). Fifty-three percent of respondents had completed secondary school. As would be expected, few businesses were full participants in the formal economy— only 10 percent of respondents reported having a tax identification number. Fifty-eight percent were sole proprietors; 24 percent had one employee. The average enterprise in the sample earned a little over U.S. $100 per month for its proprietor.

Sixty percent of the businesses we spoke to had mobiles already. Among those who have not yet purchased a mobile, most (73 percent) indicated that they were thinking of purchasing one within the next twelve months. By comparison, more than 90 percent of respondents had a television at home. Roughly 25 percent had a landline, either at home or at work. Seven percent of respondents reported having a PC; those were concentrated in the two highest socioeconomic strata. Among the enterprises with mobile connections, 96 percent had prepaid plans; 9.5 percent reported owning a GPRS-enabled or data-ready handset. Of those with such handsets, 40 percent pay for the data connection.

Against this background of prepaid, nondata-enabled handset ownership, Figure 8.1 contains the basic data for this chapter's argument; among the small and informal business segment in urban India, most use their handset to receive calls, while some apparently try to avoid making calls, presumably by engaging in missed-call behaviors.[10] *Text messages are the most popular nonvoice communicative activity, and are used by the majority of respondents.* Thirty percent are regular texters, and another 40 percent text from time to time. A smaller but sizable proportion play games; others (32 percent) listen to music. Photos, e-mail, and the Web barely register.

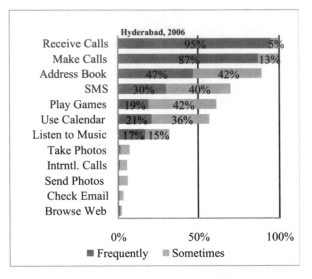

Figure 8.1   Mobile features used by small-business owners in Hyderabad, 2006.

To some extent, these patterns are a reflection of the relative importance of
communication versus information processing among these small enterprises.
While calls may be placed and received for both business and personal uses, it
is often the case that the businesses themselves require little in the way of formal information processing or information search. To run a cash-only retail
shop, a cobbler, a tailor, or a taxi or to prepare and sell food from one's home
does not require a speedy Internet connection, a database, or even an e-mail
account. However, it has proved helpful to have a mobile phone to coordinate with customers and suppliers locally and beyond. A basic telephone connection—which the mobile finally makes possible—goes a long way towards
improving the productivity of many small and informal enterprises.[11]

SMS AS MOBILE MEDIA

The central question for this chapter is now apparent; how might a small or
informal business owner use mobile media? The remainder of the chapter
must make the link, in largely hypothetical terms, between what is possible in
the varied mobile media space—described by other chapters in this volume—
and what is currently the reality of the small and informal business in the
developing world. The strongest part of that link is likely to be the SMS.

   This is not to say that all SMS messages are necessarily mobile media.
Indeed, this volume brings into question distinctions between telecommunications and media within the space of mobile technologies. The conceptual

breaking point is of particular importance to this paper, since it is dealing with the low end of users' experiences. And yet a full explication of the term *mobile media* may uncover presumed differences in form (text, music, voice, image, video), relationship (one to one, one to many, many to many), and permanence (real-time messages, asynchronous messages, stored/created content), with highly variable sets of conclusions as to what constitutes a mobile medium.[12] Indeed, we may agree only that the conventional person-to-person voice call is not a mobile medium (although it is certainly mediated communication). So too might the "traditional" person-to-person text message be considered mediated communication rather than a mobile medium. However, the characters of the text message take a variety of forms beyond single messages composed by one person and sent to another: anonymous p2p forwards of clever sayings or vicious rumors, one-to-many "broadcasts," automated status messages on everything from account balances to movie times to reminders to take one's medicine. It is these other forms, built on the humble SMS, which comprise the surprisingly potent and diverse SMS-based low-cost mobile media. In the remainder of the chapter, we'll focus on applications using the SMS and leave p2p forwards for another discussion.

Quite simply, the SMS is aligned with the needs and behaviors of the archetypal informal businesses owner, and with those other resource-constrained uses in the developing world. At the equivalent of a few U.S. cents per 160 characters, the SMS is not a particularly efficient way to move digital content from place to place. However, for many new users in the low- and even middle-income brackets, the transparency and predictability of the pricing of SMS messages makes it very easy to meter use.[13] Indeed, these pricing mechanisms have been proven successful with a range of products and services targeted for low-income users in developing countries—in many cases, resource-constrained customers are willing to pay higher per-unit costs in order to avoid upfront costs or lasting commitments.[14] Furthermore, the capability to send and receive SMS messages comes preloaded on every handset and as part of virtually every tariff plan; there are no additional hardware costs, no software costs, no upfront subscriptions to pay, and relatively few skills to learn.

Barriers to texting do exist. Tariff plans vary from place to place. In some countries, text messages are extremely inexpensive, and use is higher. In others, they are priced at less of a discount relative to voice calls, and usage is lower. There are also key issues of language and literacy. Some languages are better suited to texting than others, given the complexity of the language, the kind of script, and the availability of handset support. And, of course, users must be sufficiently literate in a supported language to craft and send, or receive and read, a text successfully.[15] However, at this point in 2008, fewer skills and financial resources are required to send a text message than to configure a mobile device to send and receive e-mail, or access the Web. Until the mobile Web is more affordable, accessible, and easy to use, it is the humble SMS which will be the de facto or default way by which a large proportion of the planet—including those archetypal small and informal businesses—sends, receives, crafts, and stores electronic data, to the tune of what some analysts predict will be 2 trillion text messages in 2008.[16]

EXAMPLES OF SMS-BASED LOW-COST
MOBILE MEDIA APPLICATIONS

The landscape of SMS-based mobile media in the developing world is already varied, and is likely to become more so in the next few years. In *form*, some systems are self-contained and stand-alone; others function as bridges to or hybrids of other media forms, particularly the Internet. In *source and mission*, some SMS-based mobile media are products of the telecommunications industry or broader marketplace; others have been developed by NGOs or other entities concerned specifically with the information and communication needs of resource-constrained users.

Stand-alone systems introduced by the industry are the earliest and likely the most familiar to readers of this chapter. Whether offered directly by the mobile operators or through third parties via short codes, subscriptions to or one-off requests for horoscopes, jokes, cricket scores, city guides, news, and other SMS premium content have been popular in India since the arrival of SMS. Real-time services like traffic alerts are also available.

More recently, providers in India and elsewhere have been introducing m-banking systems. Mobile banking is a particularly powerful example of an advanced service delivered via SMS, which may prove particularly useful for many of India's small and informal enterprises. The technology allows users to transfer and store value on accounts linked to or accessed via their handset and Subscriber Identity Module (SIM) card, and to manage payments and person-to-person transfers generally using a text-message channel. The services are hosted by telecommunications providers, often in partnership with banks. The take-up has been strongest in the Philippines and in South Africa, but experiments are underway around the world, and rollout in India is imminent. For those with no access to a PC, or a bank account, mobile banking holds the promise of a path to participation in formal, safe, reliable transactions at a distance.[17]

Other stand-alone initiatives have been launched by development agencies, governments, NGOs, or other nonprofit actors. For example, Mobile for Good (M4G), in Kenya, offers job search, health advice, and community news applications using the SMS channel. In South Africa, Cell-Life offers an application which sends SMS reminder messages to patients taking antiretroviral drugs for HIV/AIDS.[18] Whether built for profit, for social/developmental impact, or both, these stand-alone systems leverage the text message to deliver to (or gather from) users increasingly advanced, customized, or personalized kinds of information. For the hundreds of millions of people with access to mobiles but not to PCs, these systems can broaden and enrich their participation in community, national, and international information landscapes.

Other applications are more nuanced in their relationship to the existing information landscapes. A variety of SMS systems have been deployed which work in concert with other media forms to improve the portability, reach, or interactivity of broader media systems. The most prominent and widespread examples of these forms are probably the interactive SMS/TV linkages. Kraidy described the SMS votes used to power *Indian Idol* and other programs as "inter-media."[19] These systems provide ways for viewers to interact with previously one-way broadcast TV via their inexpensive handsets, and, in some sense, have succeeded where other more elaborate forms of interactive television have failed.

Equally interesting and, in the long run, perhaps more powerful are the bridges made via SMS between simple mobile phones and the Internet. Some of the large Internet search engines offer simplified searches via SMS. These are not free but nevertheless provide access to information maintained on the Web from even the most basic of handsets. Funding from or partnerships with international development agencies have helped launch companies like Ghana's Tradenet and Senegal's Manobi, multiplatform systems for b2b services and agricultural commodity exchanges. While prospective trading partners may find each other using the SMS functions, the systems also offer news and price quotes drawn from information on the broader Internet.[20]

There are other forms of low-cost mobile media which are not SMS-based. Voice applications have been applied in a variety of ways, such as the MobilED project (a form of audio-access to Wikipedia) used in education in South Africa, or the pilot of Question Box, which lets users initiate a Web search using an intercom or telephone link to live/human operator doing the search.[21] These voice interfaces help erode the barriers due to illiteracy and the limits of 160 characters, but, due to the relative costliness of voice calls and the imprecision of speech recognition software, do present challenges of their own. Nevertheless, continued experimentation applying the speech approach to low-cost mobile media is important and will undoubtedly continue.[22]

DISCUSSION

It would be unnecessary reductionist to suggest that stand-alone or hybrid systems will prevail, or are more relevant or appropriate for small and informal business owners, let alone overall. However, as we step back from the examples and consider the range of SMS-based mobile media applications already on offer in the developing world, there is a practical and theoretical appeal to those hybrid systems—the ones which bring some elements of the Internet to the basic handset via the humble SMS.

From a practical standpoint, those systems which link the Internet to the SMS may exhibit more flexibility and power. The conventional Internet—domain of plentiful big screens, full keyboards, and powerful data-transmission and storage options—offers a wealth of advantages unavailable at 160 characters at a time. So, if one were to design anything as mundane as a movie tickets purchasing application, or as sophisticated as a supply chain management system for small and informal enterprises, it is appealing to do so in a way that those who have the access and skills to use the "conventional" Internet can do so, without excluding those who cannot.

Indeed, a couple of recent systems illustrate how flexible SMS+Internet hybrid approaches can be. My colleagues at Microsoft Research India recently created a customizable SMS application to support an agricultural pricing information system for farmers. What is notable about this SMS server system is that it allows a PC, when coupled to one smartphone, to be a direct bridge between other low-end mobile phones and content stored on the PC or on the Internet. In its first implementation, farmers take advantage of the hybridity of the system to query the Web without ever directly seeing a PC, let alone using a device with an IP address. The system is low cost; farmers access it via their $30 handsets, and the agricultural collective running it does so without needing to purchase bulk SMSs or enter into contracts with mobile carriers. A similar PC-based tool kit for NGOs called frontline SMS has been used by a variety of organizations around the globe to build their own SMS applications.[23]

From a theoretical standpoint, the blurring of the Internet and the low-cost handset via the SMS challenges researchers considering the essence of mobile media. This case illustrates that the significance and use patterns of mobile media are enabled and constrained not only by what is available in the marketplace but to a very broad extent by socioeconomic factors and contextual needs. Radically different use patterns—for example, around per-use satchel purchases—exist and will persist in the developing world. The forms mobile media will take under these conditions will be different indeed. Conversely, if the Internet is only as useful as its slowest or weakest link, then we must imagine the utility and appeal of "the Internet" as presented 160 characters a time. That may be "mobile media" for some users for some time to come.

In the longer run, SMS messages may be replaced by IP-based mobile e-mail or instant messaging, as may be beginning to happen among more prosperous users in the developed world.[24] But in the meantime, the humble SMS is likely to ramble on and, indeed, remain invigorated by a set of "mobile media" additions to its offerings. In *The Shock of the Old*, David Edgerton makes a case for examining technologies in use—as opposed to the cutting edge—in the spirit of understanding "Creole" technologies like the oxcart on the urban street or the dhow sailing boat in the mechanized harbor. Generally speaking, the mobile phone, and particularly its SMS function, have some Creole attributes. They are adopted and appropriated in ways that fit existing patterns of local life and daily needs.[25] So, too, might "Creole" mobile media persist, long after innovations have removed their underlying technologies from the cutting edge.

A second theoretical issue is that of the persistent gap between those with access to the Internet and those without. The Hyderabad survey points to a broader issue—that across India and the rest of the developing world there represents at least a continuum, and probably a bifurcation, of media experiences. Cartier, Castells, and Qiu capture this bifurcation in their description of low-cost, working-class ICTs for "the information have-less" in urban China.[26] Some at the top have PCs, laptops, landlines, fancy TVs, mp3 players, and such. Others make do with a modest TV (or no TV at all) and, increasingly, with an inexpensive mobile phone kept active for a few dollars a month via a prepaid card. Mobile media systems which provide access to the Internet, even with reduced richness, may begin to erode some of the bifurcation, whereas a continued reliance on separate, stand-alone systems may do more to exacerbate it. The richness of the interactive experience may not be on par with that offered by a 3G phone or a laptop, but hybrid low-cost SMS-based mobile media offer access and some of the functionally Web and PC users take for granted. To the extent that pilots turn in to scalable projects that suit real user needs (in education, livelihoods, health, government, financial services), this is encouraging indeed.

CONCLUSION

For many Indians, the talk about converged devices, or even about traditional PCs, is fairly far off. Current choices include getting by on no connectivity at all, visiting local telephone booths, or saving some money to purchase a mobile phone. Nevertheless, the humble SMS, when used to access advanced services provided over the network, can allow even basic handsets to handle information search, processing, and storage functions. These low-cost/hybrid forms exist elsewhere in the world, but as this chapter has argued, they are of special significance in the developing world for users who lack access to alternative sources for new media applications and content on PCs and the Internet.

Whether stand-alone or hybrid, the low-cost SMS-based mobile media will persist and be useful. Hybrid forms especially may prove to be powerful, flexible ways to extend the reach and usefulness of the Internet itself. It is an open question as to how many of the archetypal small and informal firms will adopt such systems in the future, but some almost surely will. As long as lower-income users in the developing world have the opportunity to use it, instead of higher-priced or more complex alternatives, the SMS will remain, resilient and persistent, longer than we might imagine.

NOTES

1. Tarmo Virki, "Nokia's Cheap Phone Tops Electronics Chart," Reuters (May 3, 2007), www.reuters.com/article/companyNewsAndPR/idUSL0262945620070503?sp=true; United Nations Conference on Trade and Development, *Information Economy Report 2007–2008: Science and Technology for Development—the New Paradigm of ICT* (Geneva: UNCTAD, 2008).

2. Telecom Regulatory Authority of India, "The Telecom Services Performance Indicators July–September 2007" (New Delhi: TRAI, 2008), hereafter cited in text.

3. Pippa Norris, "Digital Divide: Civic Engagement, Information Poverty, and the Internet Worldwide," in *Communication, Society, and Politics*, ed. W. Lance Bennett and Robert M. Entman (New York: Cambridge University Press, 2001); Manuel Castells, *The Rise of the Network Society* (Malden, MA: Blackwell, 1996).

4. Anandam Kavoori and Kalyani Chadha, "The Cell Phone as a Cultural Technology: Lessons from the Indian Case," in *The Cell Phone Reader: Essays in Social Transformation*, ed. Anandam Kavoori and Noah Arceneaux (New York: Peter Lang, 2006).

5. Jonathan Donner, "Research Approaches to Mobile Use in the Developing World: A Review of the Literature," *The Information Society*, 24 (2008): 140–59; Manuel Castells, Jack Linchuan Qiu, Mireia Fernández-Ardèvol, and Araba Sey, *Mobile Communication and Society: A Global Perspective (Information Revolution and Global Politics)* (Cambridge, MA: MIT Press, 2007).

6. Donald C. Mead and Carl Leidholm, "The Dynamics of Micro and Small Enterprises in Developing Countries," *World Development*, 26 (1998): 61–74.

7. Previous studies suggest that the telephone is often more important than the PC. Earlier work focused on the landline. Richard Duncombe and Richard Heeks, "Information, ICTs and Small Enterprise: Findings from Botswana," in *Small Scale Enterprises in Developing and Transitional Economies*, ed. Homi Katrak and Roger Strange (New York: Palgrave, 2002), 285–304. Recent explorations contrast the mobile and the PC, Steve Esselaar, Christoph Stork, Ali Ndiwalana, and Mariama Deen-Swarra, "ICT Usage and Its Impact on Profitability of SMEs in 13 African Countries," paper presented at the *International Conference on Information and Communication Technologies and Development*, Berkeley, CA (25–26 May 2006). Thomas S. J. Molony, *Food, Carvings and Shelter: The Adoption and Appropriation of Information and Communication Technologies in Tanzanian Micro and Small Enterprises* (dissertation, University of Edinburgh, 2005).

8. Jonathan Donner, "Internet Use (and Non-Use) among Urban Microenterprises in the Developing World: An Update from India," paper presented at the *Conference of the Association of Internet Researchers*, Brisbane, Australia, September 28–30, 2006.

9. Find details on the Hansa quintles at http://www.hansaresearch.com/IRS_2007_Brochure.pdf. Trained enumerators conducted face-to-face intercepts in both home and business settings in both municipalities. Ideally, we would have gathered a purely random sample, but as is often the case in developing countries, no comprehensive list of informal businesses was available.

10. Jonathan Donner, "The Rules of Beeping: Exchanging Messages via Intentional "Missed Calls" on Mobile Phones," *Journal of Computer-Mediated Communication,* 13(1) (2007), http://jcmc.indiana.edu/vol13/issue1/donner.html.
11. See note 7. Also, Jonathan Donner, "The Use of Mobile Phones by Microentrepreneurs in Kigali, Rwanda: Changes to Social and Business Networks," *Information Technologies and International Development,* 3(2) (2006): 3–19.
12. The lines between mass and interpersonal communication have never been clear; mediated communication is problematic, relative to both. For example, see Steven H. Chaffee, "Mass Media and Interpersonal Channels: Competitive, Convergent, or Complementary?" in *Inter/Media: Interpersonal Communication in a Media World,* ed. Gary Gumpert and Robert Cathcart (New York: Oxford University Press, 1982).
13. Sonja Oestmann, *Mobile Operators: Their Contribution to Universal Service and Public Access* (Intelecon Research & Consultancy: 2003, http://rru.worldbank.org/Documents/PapersLinks/Mobile_operators.pdf).
14. Ayesha Zainudeen, Rohan Samarajiva, and Ayoma Abeysuriya, "Telecom Use on a Shoestring: Strategic Use of Telecom Services by the Financially Constrained in South Asia" (Colombo, Sri Lanka: LIRNEasia, 2006). C. K. Prahalad, *The Fortune at the Bottom of the Pyramid: Eradicating Poverty through Profits* (Upper Saddle River, NJ: Wharton School Publishing, 2005).
15. Jan Chipchase, "Understanding Non-Literacy as a Barrier to Mobile Phone Communication" (Nokia Research Center, 2005, http://research.nokia.com/bluesky/non-literacy-001–2005/index.html).
16. Research and Markets, "2008 Global Mobile—Data and Content Markets" (February 2008, http://www.researchandmarkets.com/reports/c84203).
17. infoDEV, "Micro-Payment Systems and Their Application to Mobile Networks" (Washington, DC: InfoDEV, 2006; http://infodev.org/files/3014_file_infoDev.Report_m_Commerce_January.2006.pdf).
18. Organizational Web sites at http://uk.oneworld.net/section/mobile; http://www.cell-life.org/.
19. Marwan Kraidy, "Inter-Media Dynamics and Reality Television in the Arab Region," paper presented at the *Command Lines: The Emergence of Governance in Global Cyberspace,* University of Wisconsin–Milwaukee, April 29–30, 2005.
20. Organizational websites at www.tradenet.biz; www.manobi.sn/sites/za/
21. Merryl Ford and Adele Botha. "MobilED—An Accessible Mobile Learning Platform for Africa?" paper presented at the *IST Africa 2007 conference,* Maputo, Mozambique, May 2007. Questionbox organization Web site at http://www.questionbox.org/
22. For more on voice approaches to hybrid media, see Nathan Eagle, "The Mobile Web is NOT Helping the Developing World . . . and What We Can Do about It," Mobileactive.org, 5 December, 2007, http://mobileactive.org/mobile-web-not-helping-developing-world-and-what-we-can-do-about-it-guest-writer-nathan-eagle.
23. Rajesh Veeraraghavan, Rajesh, Naga Yasodhar, and Kentaro Toyama, "Warana Unwired: Mobile Phones Replacing PCs in a Rural Sugarcane Cooperative," paper presented at the *2nd IEEE/ACM International Conference on Information and Communication Technologies and Development* (ICTD2007), Bangalore, India, 2007. Frontline SMS organizational Web site http://www.frontlinesms.com/.
24. Asher Moses, "For SMS, the Days Are Numbered," *The Age,* July 27, 2007, http://www.theage.com.au/news/mobiles—handhelds/for-sms-the-days-are-numbered/2007/07/27/1185339221496.html.

25. François Bar, Francis Pisani, and Matthew Weber, "Mobile Technology Appropriation in a Distant Mirror: Baroque Infiltration, Creolization and Cannibalism," paper presented at the *DIY speaker series* at the Annenberg Center for Communication at USC, Los Angeles, CA, April 7, 2007. David Edgerton, *The Shock of the Old: Technology and Global History since 1900* (New York: Oxford University Press, 2007).
26. Carolyn Cartier, Manuel Castells, and Jack Linchuan Qiu, "The Information Have-Less: Inequality, Mobility, and Translocal Networks in Chinese Cities," *Studies in Comparative International Development,* 40 (2005): 9–34.

9 Innovations at the Edge
The Impact of Mobile Technologies on the Character of the Internet

Harmeet Sawhney

INTRODUCTION

New communication networks typically start as complementary systems that extend the reach of an established network to areas it could not penetrate. For instance, people had to walk to a telegraph office to send a telegraph and conversely a delivery person had to deliver a telegraph to the home of the recipient. The telephone extended the reach of the telegraph to the homes of individual customers and thereby eliminated the need for hand delivery of messages. Similarly, cellular networks extended the reach of wireline networks to the places they could not reach—moving vehicles. Today, we are witnessing the growth of mobile networks that extend the reach of the Internet to mobile environments. However, the similarity ends there in one very significant way.

In the past, the new technology came into play after the institutional arrangements for old technology had settled. The expansion of telegraph networks in the United States was chaotic until Western Union rationalized them. So when Bell started developing the telephone, he had to deal with an entrenched old network. Later, the cellular technology grew in the shadows of an entrenched Bell System, which by then was the dominant communications company. However, mobile technology has come into play while the Internet is still an unsettled technology.

The freewheeling character of the Internet has been under the assault of the rationalizing forces ever since advent of e-commerce. It remains to be seen whether the Internet we know is only a "democratic moment" or will the libertarian forces prevail in keeping the Internet relatively open. It is an unsettled matter. Within this context, the advent of mobile technologies has added another dimension to the interplay of forces shaping the character of the Internet. On the one hand, we see the carryover of the libertarian impulse onto the bottom-up wifi networks. On the other hand, we see technologies such as 3G, iMode, and iPod serving as vehicles for the introduction of the top-down logic at the edges of the Internet.

"Chaos" and "order" have tended to be oppositional configurations. The Roman roads were emblematic of the imperial order. But at the outer edges of the Roman roads, the frontiers of the imperium, the barbarians lurked. For most of Roman history, the advancing empire kept the barbarians on the run. But when the empire started losing energy, the barbarians overran Rome. The Chinese emperors built the Great Wall of China to keep the nettlesome barbarians out. Eventually, the barbarians overran the imperial order. In the realm of communications, for a period of time there was the uneasy coexistence of radio amateurs, who developed broadcasting, and the radio corporations. Eventually, the corporate order prevailed and the amateurs were marginalized. Thus, we see that, on the one hand, "chaos" seeks to break down "order" and, on the other hand, "order" seeks to stamp out "chaos." Both seek the elimination of the other.

Within this context, the interface between the chaotic Internet and its mobile extensions, which by and large are ordered configurations, begs attention. For the purposes of this particular paper, the focus will be on the following two questions: Will the chaos of the Internet spill over and open up the wireless arena? Or will the order of the wireless arena march onto the Internet?

The recent developments with regard to iPhone highlight these tensions. In the run-up to the launch of iPhone, Steve Jobs said,

> We define everything that is on the phone. You don't want your phone to be like a PC. The last thing you want is to have loaded three apps on your phone and then you go to make a call and it doesn't work anymore. These are more like iPods than they are like computers . . . These are devices that need to work, and you can't do that if you load any software on them. That doesn't mean there's not going to be software to buy that you can load on them coming from us. It doesn't mean we have to write it all, but it means it has to be more of a controlled environment.[1]

Coming from Steve Jobs, talk of a "controlled environment" is ironic because he was the upstart who disrupted the established hierarchical paradigm of mainframe-based computing with the launch of Apple personal computers. His company's stance was nicely captured in the still-talked-about "1984" commercial in which a runner throws a hammer and brings the big brother down, the former representing the personal computer and latter the mainframe. Perhaps, now as a corporate mogul, his perspective has changed.

A charitable view, from a radical point of view, is that right now Jobs has no choice but to present a controlled environment perspective because otherwise AT&T, the service provider that provides connectivity to iPhone, would not partner with Apple. According to this theory, later when iPhone becomes a major success and has market power, Apple will dictate terms to the carriers and open up the architecture of iPhone.[2] One supporter of this view thinks that

> . . . if the iPhone becomes a hit over the next year or two, Apple will get more and more leverage over the Carriers and won't be afraid to use it.

While we wish OS X could be used to its full potential for us, and we wish OS X could be used as a Trojan horse for the masses, maybe the iPhone is itself a Trojan horse for the whole mobile phone industry.[3]

If that does happen, it would be ultimate subversion of order. More realistically, this is perhaps wishful thinking because Job is not only celebrating the closed environment of iPhone, which he had to according to this theory, but also that of iPod, which he did not have to.

Notwithstanding what Steve Jobs says and does, there are other players for whom closed environments are anathema. Within three weeks of the launch of iPhone on June 29, 2007, there were reports of a hacker, who goes by the name "Nightwatch," having created and started an unauthorized program on iPhone. On August 18, 2007, there was the following post on mobile-society@googlegroups.com listserv:

> The group iPhone Dev Wiki has discovered a way to partially unlock the device so that it can work with any AT&T or Cingular SIM card without the need for the exclusive new contract . . . If it's good news, it needs to be spread!![4]

In the context of our discussion, the last line is notable. The hackers are particularly focused on decoupling iPhone from its sole authorized cellular service provider, AT&T.[5] Interestingly, this is reminiscent of hackers unlocking Apple's original restriction that an iPod could share music with a single Mac and also it could not be used with a Windows PC.[6] As Aviel D. Rubin of Independent Security Evaluators, a company that tests the security of its clients' systems by hacking them, says, "Anything as complex as a computer—which is what this phone is—is going to have vulnerabilities."[7]

> The hackers smell blood.
> It remains to be seen whether chaos or order will prevail.

This paper employs the "arena of innovations" framework developed by Sawhney and Lee[8] to analyze how the juxtaposition of chaos and order vis-à-vis Internet and its mobile extensions will play out. It uses iPhone as an example, among others, to draw insight and acuity. The next section provides an overview of the arena of innovation framework. The subsequent sections then employ this framework to study the development of the mobile extensions of the Internet.

ARENAS OF INNOVATION[9]

In the Sawhney and Lee paper we first discussed the case of broadcasting to provide a vivid example of an arena of innovation in action. Thereafter we discussed the structural characteristics of an "arena of innovation" that allows mass of loosely organized amateurs to generate insights that escape corporate laboratories. As we will later see, the arena of innovation framework is a useful heuristic for analyzing the intersections between mobile and Internet technologies.

In the early stages of radio's development it was framed as a wireless version of the existing telegraph technology. In fact the entire discourse was based on the telegraph metaphor. It was seen as a technology that extended the reach of the wire-based telegraph network into difficult-to-connect places such as ships in mid-sea. Even Guglielmo Marconi, the inventor of the radio, focused all his energies on molding radio into a technology for point-to-point communication. So much so, he saw the tendency of radio waves to scatter as a major nuisance.[10]

The influence of the telegraph metaphor persisted for a long time. At the time of the GE-AT&T-Westinghouse agreement,[11] the institutional forces guided by the telegraph analogy were clearly working towards casting radio as a point-to-point technology. In fact the very thought of broadcasting was not even within the realm of imagination.[12] It was the stubborn refusal of renegade amateurs to comply with the larger institutional framework that resulted in the identification of broadcasting as a new means of communication. Frank Conrad, in early 1920, started transmitting phonograph music as part of his ongoing experiments over a radio transmitter. His signal was picked up by amateur radio buffs and their enthusiastic response led Conrad to schedule regular concerts, which attracted much newspaper coverage and publicity.

Harry P. Davis, a Westinghouse vice-president, soon realized that "efforts then being directed to develop radiotelephony as a confidential means of communication were wrong, and that this field instead offered one of widespread collective publicity."[13] He authorized the construction of a Westinghouse transmitter in East Pittsburgh and the transmission of programs on a regular basis. This transmitter—which was assigned the historic call letters KDKA—went on air on November 2, 1920, and it marked the birth of broadcasting.[14]

The audiences grew at a phenomenal rate and broadcasting became a big business. People in the wireless industry could not help but wonder how come they did not see something as obvious as the potential for broadcasting. William C. White, a scientist at the General Electric research laboratory, later recalled, "[I was] amazed at our blindness . . . we had everything except the idea."[15] The problem was with the conceptual templates that were brought to bear upon the phenomenon. The telegraph metaphor guided all the speculative activity along the point-to-point dimension and blinded people to any new configurational potentialities opened up by the new technology.

Such examples abound on the Internet, which is different from radio and other communication technologies in that it offers not one but many new configurational potentialities. On the one hand, amateurs have sought to subvert dominant institutions of the off-line world by replicating them on the Internet, for example, Internet radio and telephony. On the other hand, they have developed new configurational capacities that have forced established institutions to change, for example, Napster and other peer-to-peer systems and Web sites like YouTube and MySpace. It is perhaps because of the protean nature of the Internet that the rationalizing forces of commerce have not yet been able to quash the libertarian culture on the Internet. It is also perhaps in the continual development by amateurs of new and destabilizing configurational capabilities the libertarian hope for the Internet's future lies.

After studying the above-discussed cases, we then identified the following defining characteristics of an arena of innovation:

PHYSICAL SUBSTRUCTURE

1. It is constituted by a multitude of people.

   By its very nature, an arena of innovation requires the participation of a very large number of people. It cannot be created by a handful of people, no matter how determined. Why is size important? The power of a mindful multitude is very nicely captured by Eric Raymond, who when explaining the success of open source movement, states, "given enough eyeballs, all bugs are shallow."[16]

2. The participants are dispersed across space.

   In the case of arenas of innovation, the geographical dispersion of the participants is critical. For example, if all the radio amateurs were collected together in one place, they would not form an arena of innovation. The radio amateurs could spot the potential of broadcasting because they were dispersed across space. Since the new configurational potentiality of a communication technology by its very nature is a new modality of communication across space, geographical dispersion of participants is essential for its identification.

3. The object of play of the participants is the very medium that interconnects them.

   The size provides numerous "eyeballs," each differently oriented, and the geographical dispersion creates the arena within which a new configurational potentiality is spotted. These two factors define the overall arrangement of the arena. The third physical element is related to the phenomenon that is observed. The participants play around with the same communication technology that ties them into a geographically dispersed community. It is in the course of this fun and play that a new configurational potentiality is identified.

SOCIAL SUBSTRUCTURE

1. The experimentation by the participants is driven by fun rather than commercial gain.

   The fun is primarily in the act of experimentation itself and not the results it may or may not produce. While leisure activities often generate learning, it is basically a happy by-product of a fun activity. The clever but goalless activity in this milieu energizes and greatly enhances the chance that one of the mindful "eyeballs" among the thousands that come together to create the arena of innovation will spot a new configurational potentiality. Corporate research teams, on the other hand, are boxed in by goals. Furthermore, there is constant pressure to show results. These factors retard the possibility of out-of-the-box thinking.

2. The barriers to entry for joining the community are low.

   The low barriers to entry allow all kinds of individuals to enter the arena and do their own thing and thereby enhance the possibility of a breakthrough observation or idea.

3. There is camaraderie and openness that facilitate cross-fertilization of ideas within the community.

   In leisure groups communication is much freer than in commercial settings because the ties between members are based on the "principle of reciprocity" or "the gift relationship." While members compete with each other and seek to establish reputations, they also help each other and freely share their knowledge. Each time a member helps another member he or she can be quite confident that somewhere down the line somebody else will provide assistance when it is needed. This free flow of information greatly enriches the overall environment.

MOBILE EXTENSIONS OF THE INTERNET

With the arena of innovation as the backdrop, we will now look at the mobile extensions of the Internet. We will focus on MP3 players and cellular phones because they seem to be evolving towards open interconnected networks, potential technological underpinnings of an arena of innovation, from two opposite directions. The MP3 players,[17] stand-alone devices, are starting to get connected. Cell phones, connectivity technologies, could be moving out of a closed system architecture to an open one.

I have been watching the spread of MP3 players with great interest ever since the publication of the arenas of innovation paper.[18] I anticipated that soon they will start getting interconnected and we may see the emergence of a new arena of innovation. My expectation that the MP3 players will start getting interconnected was based on a model I had earlier developed, based on research on clocks and computers, that delineates stages in the decentralization of a technology—over time there is a reduction in the size of a technology, multiplication in its numbers, and diffusion into the social fabric and eventually interconnection.[19] The MP3 players were in the diffusion stage and the model suggested some kind of interconnection down the line.

Eventually, my anticipation did pan out in the form of Microsoft's Zune, which offers connectivity with other Zune's within thirty feet. The connecting Zunes can transfer audio files, playlists, and pictures. But there is a "three-plays-in-three-days" limitation, that is, a transferred music file can only be played thrice within three days. Also, recipients cannot transfer files to others and not all Zune Marketplace downloads are distributable.

Microsoft celebrates the social nature of Zune. Its tagline for Zune is "the social." The package proclaims "welcome to the social" and the product logo is a network mnemonic. According to Microsoft,

> Under the Zune brand, Microsoft will build a community for connecting with others to discover new music and entertainment . . .We see a great opportunity to bring together technology and community to allow consumers to explore and discover music together.[20]

"Community" has lately become a sexy word in the business world, as in other spheres.[21] User communities, when nurtured properly, offer the prospects of harvesting free user inputs with regard to product design and marketing at minimal cost. From a business point of view, "the community is best described, as an organization where consumers participate in costless assets formation by playing with a product."[22] The spectacular successes of various social networking systems on the Internet seem to have fueled this appetite. In this vein, Microsoft wants a community but with the various limitations noted earlier. In effect, it wants a controlled community, controlled in various ways that keep its assets secure and maximize profit potential.

We know what Microsoft wants. The question is can Microsoft have what it wants.

Carolyn Marvin points to the difficulties inherent in such wishes:

> New media embody the possibility that accustomed orders are in jeopardy, since communication is a particular kind of interaction that actively seeks variety. No matter how firmly custom or instrumentality may appear to organize or contain it, it carries the seeds of its own subversion.[23]

These two pithy and sagacious sentences are worthy of elaboration into an essay, something this author hopes to do down the line. For our current purposes, we will focus on the elements most relevant for our present discussion.

In the case of Zune, Microsoft is trying to maintain control by "instrumen-tality" or design that "organizes" and "contains" interactions. In effect, it is trying to impose limitations on types of interactions by channeling behavior along desired paths. But then, as Marvin tells us, once a medium opens up new ways of communication, people are almost certainly going to experi-ment endlessly or "actively seek variety" and generate surprising discoveries and innovations. Sooner or later they are likely to break out of the imposed limitations and subvert the established order. The chances are particularly high for a device like Zune, which invites play and that too on a continual basis since people carry it around during large part of their waking hours.

Already people have been hacking away at iPod, which the market leader Zune is trying to overtake. On the one hand, the hackers have sought to remove those features of the iPod they find annoying. For example, Engad-get helps people replace the "No" symbol that appears when syncing iPods, which many people find annoying, with a more amiable substitute. On the other hand, hackers have sought to add new capabilities to iPods. For example, iPodMAME allows people to play the Pac-Man game on their iPods. The main thrust of the hackers has been to run Linux on iPods, which provides the platform for myriad applications.

In many ways, Microsoft has conceded inability to control users by agreeing to pay Universal Music $1 for every $250 Zune.[24] This arrange-ment is unprecedented because in the past music companies have been paid per song and not a slice of the cost of hardware. The royalty Microsoft pays Universal Music per piece of hardware has been dubbed the "piracy tax" because it is based on the assumption that the buyers will inevitably use the device for acquiring, storing, and playing unauthorized music.[25] According to Hollywood mogul David Geffen:

> . . . each of these devices is used to store unpaid-for material. This way, on top of the material people do pay for, the record companies are get-ting paid on the devices storing the copied music.[26]

Similarly, the *New York Times* reported that Universal Music character-ized royalty per device as "only fair" because they were likely to be "reposi-tories for stolen music." This stance is not unreasonable considering the fact that a recent study estimated that Apple sold only about twenty songs per iPod, a small fraction of the device's carrying capacity. The great bulk of the music files on iPods are acquired from ripped CDs, personally owned disks, and the Internet.[27]

Software piracy has been a long-standing issue. The decreasing costs of computer peripherals have facilitated increasing levels of piracy.[28] Another contributing factor has been easy access to Internet bandwidth.[29] What would have earlier taken days to download, via dial-up connection, now takes only minutes over broadband networks. Connectivity, especially broadband, increases opportunities and temptations for piracy.[30]

In the cell-phone world, where connectivity is central, we have the hackers trying to open up the architecture of the cell-phone service, which thus far has been very closed or highly integrated. The service providers dictate the parameters within which the cell-phone network is used and also the end-user device itself. We now have a Homebrew Mobile Phone Club, modeled on the original Steve Jobs and Steve Wozniak's Homebrew Computer Club, busy exploring the possibilities of opening up the cell phone. The Homebrew Mobile Phone Club's Web site describes itself as "a physical and virtual club to do to mobile phones what Jobs and Woz did to computers." There is also a Tuxphone Project, named after the penguin mascot (Tux) of the Linux movement, which states that its "objective is to create an open (in every sense of the word) cellphone platform that is convenient for creating novel applications." Other projects include OpenCell, started by students of Florida Institute of Technology.

For open-cell-phone enthusiasts, present-day cell phones are like mainframes. The users have little flexibility. They have to function pretty much within the tight parameters set by the system design. The open-cell-phone enthusiasts hope to do to this paradigm what the Jobs and Woz generation did to the mainframes. In effect, they want to make the cell phones as flexible as the personal computers.[31]

Unlike "homebrew" personal computers, the ambitions of "homebrew" cell-phone projects are not limited to the device itself but go beyond to the connectivity between them. The "homebrew" mobiles are vehicles for opening up the cellular network. As Suraj Patel, the force behind the Tuxphone Project, says, "I want the phone to be much more open. The world's best research and development lab is all the hackers out there. Enable them, and they'll do it."[32]

Patel notes that five years ago it would not have been possible to build a cell phone without a "million-dollar" lab. Even today, it is quite a taxing affair to put together a "homebrew" cell phone. Yet the final product is clunky and the battery's charge lasts less than half an hour. "So what's the motive? For many, it is the urge to play around with devices that are an integral part of all our lives." Patel goes on to say, "The real reason you should be interested in this technology is the applications you can't predict." Similarly, Casey Halverson, a Seattle-based open-phones enthusiast, thinks as more people use mobile devices, "they will be running into more and more limits with closed systems." The hacking activities will go beyond the geeks to the general public and the result could be an open platform.[33]

All this sounds so much like the arena of innovation discussed earlier. Imagine an arena of innovation based on portable devices that are handily available most hours of the day and invite play. It could unleash tremendous amounts of creativity.

On the other hand, we need to also keep an eye on corporate efforts to harness open-source energy for their own benefit. For instance, Nokia released the source code of its mobile phone Web browser to spur the industry to move towards a single standardized Web browser. Opera Software ASA's CTO does not think that Nokia's move will have much impact because Nokia released only a small portion of the code and kept the rest proprietary. Also, the innovations in the released code are not likely to be of use to programmers working on mobile platforms other than Nokia's own S60 phones. According to him, "What I'm seeing is they're flirting with open source and trying to get the open source community interested in their platform, but it's more of a marketing thing rather than a real technical contribution."[34] Other such corporate efforts, among many others, include France Telecom's support of GPE Palmtop Environment[35] and Funambol's initiative to develop applications for Google Android with the help of open source community.[36] De Laat, after a comprehensive study of the different ways by which corporations have tried to leverage open-source efforts, concluded "that over time the open source-inspired networks developed by these companies gradually came to resemble classical corporate networks."[37] I cannot see anything special in mobile technology that will change this dynamic. Actually, as described previously, we see this same tendency in Nokia's Web browser project. Therefore, to this observer, it seems that if something dramatic comes about in the open source arena, it will at the amateur and not the corporate end.

MIGRATION BETWEEN ARENAS

If an arena of innovation supported by handheld devices were indeed to emerge, what would be its relationship to the arena of innovation supported by the Internet? Would it be an extension of the Internet or would it be different? What innovations will arise? Which way will they flow?

Currently, there seems to be more migration from the Internet to mobile, which should not be surprising since the Internet has been a hotbed of innovation for quite some time. Some of the innovation migration from the Internet to the mobiles include: efforts to build gateways between cell phones and Skype, linking of mobile's address book with social applications like LinkedIn or MySpace, enhanced mobile address books that work like IM (Instant Messenger), use of avatars on mobiles. This point need not be belabored.

We are also beginning to see the possibility of migrations the other way around. Unlike the Internet, where the written word dominates the Internet, the mobile phone environment is speech oriented. This difference in orientation is showing up in the development of avatars for mobile phones. The developers of avatars for mobile phones are working to create increasingly sophisticated avatars that use voice, as opposed to speech bubbles and text on the Internet. We are likely to soon see mobile-phone avatars whose lip movements and facial expressions are synchronized with real-time speech inputs.[38] And, more importantly for our analysis, "these developments will in turn flow back into virtual worlds."[39]

Speaking avatars could be fun. But then, there could also be migrations of the type that strengthen the forces of order on the Internet. For instance, the concept of micropayments employed with considerable success by DoCoMo in i-Mode has now been appropriated by many national broadband development projects. This is the scary part if one is of libertarian persuasion, at least on the Internet. Interestingly, in a very insightful article, Gerard Goggin and Christina Spurgeon examine how the premium rate culture (paid information services) is transferring from traditional telephony to the mobile environment and potentially limiting the scope of mobile interactivity.[40] According to them, "the mobile Internet is being redesigned most effectively and pervasively not in the open and interoperable spaces of the fixed Internet . . . but in the proprietary premium rate spaces of digital, and increasingly mobile, telecommunications."[41] Here we have a scenario where the order of the mobile environment is further reinforced by the transfer of elements from traditional telephony, which has been a bastion of order for over a century.

Only time will tell which scenario will prevail. But the possibility of a new mobile-based arena of innovation should not be discounted as fancy imagination. Surprise usually springs out of chaos and not order.

NOTES

1. http://www.43folders.com/2007/01/12/no-iphone-apps/.
2. Andy Warwick, Internet posting, January 11, 2007, 11:25, http://www.43folders.com/2007/01/11/osx-app-developers-iphone/.
3. Internet posting by Jack (last name not provided) on January 11, 2007, http://www.43folders.com/2007/01/11/osx-app-developers-iphone/.
4. "iPhone now works on Cingular SIM cards as well," http://www.mail-archive.com/mobile-society@googlegroups.com/msg00323.html. The reference in this post is to a similar post on *TechnoWiki*, July 20, 2007, http://technowiki.blogspot.com/2007/07/iphone-now-works-on-cingular-sim-cards.html.
5. John Schwartz, "IPhone flaw lets hackers take over, security firm says," *New York Times*, Monday, July 23, 2007, C4.
6. Chris Taylor, "How to Hack an iPod," *Time* magazine, Sunday, April 14, 2002, http://www.time.com/time/magazine/article/0,9171,1101020422-230383,00.html.
7. Quoted in Schwartz, "IPhone flaw lets hackers take over."
8. Harmeet Sawhney and Seungwhan Lee, "Arenas of Innovation: Understanding New Configurational Potentialities of Communication Technologies," *Media, Culture & Society*, 27(3) (2005): 391–414.
9. The second, third, and fourth paragraphs in this subsection have been extracted from Harmeet Sawhney, "Information Superhighway: Metaphors as Midwives," *Media, Culture & Society*, 18(2) (1996): 291–314.
10. Susan J. Douglas, "Amateur Operators and American Broadcasting: Shaping the Future of Radio," in *Imagining Tomorrow: History, Technology, and the American Future*, ed. J. J. Corn (Cambridge, MA: MIT Press, 1986), 35–57.

11. The intense rivalry between AT&T, General Electric, and Westinghouse led to a situation in which none of them could commercialize radio because the patents for key components were controlled by different corporations. The resulting impasse was resolved in 1920–21 via a patent-sharing agreement between the three corporations. Each corporation had access to a pool of about 1,200 patents but their areas of operations were restricted to specific applications. Very broadly, AT&T could use any of these patents for applications related to the public telephone network while General Electric and Westinghouse could use them for private networks and amateur markets. The interesting thing about this agreement is that it was based on the notion that radio is essentially a point-to-point technology. Therefore, no provision whatsoever was made for the emergence of broadcasting. The agreement collapsed once broadcasting became a reality. Each rival claimed that broadcasting fell within the area earmarked for it. The agreement was eventually renegotiated in 1926. As a result of the second agreement, AT&T decided to quit broadcasting in lieu of financial compensation and guarantees safeguarding its monopoly over the public telephone network. On the other hand, General Electric and Westinghouse were allowed to dominate broadcasting. See Gerald W. Brock, *The Telecommunications Industry: The Dynamics of Market Structure* (Cambridge, MA: Harvard University Press, 1981).

12. Brock, *The Telecommunications Industry.*

13. Harry P. Davis, "American Beginnings," in *Radio and Its Future*, ed. Martin Codel (New York: Harper & Brothers, 1930), 3–11, 6.

14. Davis, "American Beginnings."

15. Erik Barnouw, *A Tower in Babel* (New York: Oxford University Press, 1996), 73–74.

16. Eric S. Raymond, *The Cathedral and the Bazaar: Musings on Linux and Open Source by an Accidental Revolutionary* (Sebastopol, CA: O'Reilly, 1999), 41. In other words, when thousands of eager volunteers are involved in the development of a program, the bugs are more likely to be spotted than by a corporate team specially assembled for that task. Open-source movement's success stems from the fact that it is able to engage a huge number of "eyeballs" each of which "approaches the task of bug characterization with a slightly different perceptual set and analytical toolkit, a different angle on the problem" (Raymond, 41).

17. They are extensions of the Internet in that they plug into the Internet and download music and other programming.

18. Sawhney and Lee, "Arenas of Innovation."

19. Sawhney, "Information Superhighway: Metaphors as Midwives."

20. Marshall Kirkpatrick, "On Universal Music Group's Zune Tax TechCrunch," November 9, 2006, Available at: http://www.techcrunch.com/2006/11/09/on-universal-music-groups-zune-tax/.

21. For an insightful overview and critique, see Nikolaos Tzokas and Michael Saren, "Building Relationship Platforms in Consumer Markets: A Value Chain Approach," *Journal of Strategic Marketing*, 5 (1997): 105–20.

22. Lars. B. Jeppesen, *Organizing and Tapping Consumer Communities* (Copenhagen: Working Paper, Copenhagen School of Business, 2001), 23.

23. Carolyn Marvin, *When Old Technologies Were New* (New York: Oxford University Press, 1988), 8.

24. In addition to the royalty for each piece of hardware, Universal Music will also receive a portion of the revenue per song downloaded. Jeff Leeds, "Microsoft Strikes Deal for Music," *New York Times*, November 9, 2006, http://www.nytimes.com/2006/11/09/technology/09music.html?ex=1320728400&en=b3808e4590e6aa42&ei=5088&partner=rssnyt&emc=rss.

25. Many countries have tried to levy an "iPod" tax to compensate music companies for losses due to piracy. In Canada, such a tax levied as much $25 per iPod but the courts ruled it to be invalid. What is different about the Microsoft-Universal Service agreement is that it is voluntary (Kirkpatrick 2006). Marshall Kirkpatrick, "On Universal Music Group's Zune Tax," *TechCrunch*, November 9, 2006, http://www.techcrunch.com/2006/11/09/on-universal-music-groups-zune-tax/.

26. Kirkpatrick, "On Universal Music Group's Zune Tax."

27. Leeds, "Microsoft Strikes Deal for Music."

28. Samir Hinduja, "Correlates of Internet Software Privacy," *Journal of Contemporary Criminal Justice*, 17(4) (2001): 369–82, cited in Clyde W. Holsapple, Deepak Iyengar, Haihao Jin, and Shashank Rao, "Parameters for Software Privacy Research," *The Information Society*, 24(2) (forthcoming).

29. A. Graham Peace, Dennis F. Galletta, and James Y. L. Thong, "Software Piracy in the Workplace: A Model and Empirical Test," *Journal of Management Information Systems*, 20(1), 2003: 153–77, cited in Holsapple et al., "Parameters for Software Privacy Research."

30. Paul Craig and Mark Burnett, *Software Piracy Exposed* (Rockland, MA: Syngress), cited in Holsapple et al. (2008), "Parameters for Software Privacy Research."

31. John Borland, "Build-It-Yourself Cell Phones: Frustrated at Limitations on Mainstream Mobile Phones, 'Homebrew' Enthusiasts Are Building Their Own," *CNET News.com*, November 15, 2005, http://news.com.com/Build-it-yourself+cell+phones/2100–1008_3–5953682.html.

32. Borland, "Build-It-Yourself Cell Phones."

33. Quoted in Borland, "Build-It-Yourself Cell Phones."

34. Nancy Gohring, "Nokia to Open Source Its Mobile Browser Code," *IDG News Service*, May 24, 2006, http://www.networkworld.com/news/2006/052406-nokia-open-source-mobile-browser.html.

35. Paul McDougall, "Open Source Mobile Phone on the Way," *Dr. Dobb's Portal*, February, 8, 2007, http://www.ddj.com/linux-open-source/197004438?cid=RSSfeed_DDJ_All.

36. Brad Reed, "Funambol to Develop Open Source Mobile Messaging App for Android," *Network World*, November 19, 2007, http://www.networkworld.com/news/2007/111907-funambol.html.

37. Paul B. de Laat, "Evolution of Open Source Networks in Industry," *The Information Society*, 20(4) (2004): 291–99.

38. Kathy Cleland, "Face to Face: Avatars and Mobile Identities," in *Mobile Media 2007: Proceedings of an International Conference on Social and Cultural Aspects of Mobile Phones, Convergent Media and Wireless Technologies*, ed. Gerard Goggin and Larissa Hjorth (Sydney: University of Sydney, 2007), 33–46. See also Cleland, this volume.

39. Cleland, "Face to Face," 39.

40. Gerard Goggin and Christina Spurgeon, "Premium Rate Culture: The New Business of Mobile Interactivity," *New Media & Society*, 9(5) (2007): 753–70.

41. Goggin and Spurgeon, "Premium Rate Culture," 764–65.

10 Media Contents in Mobiles
Comparing Video, Audio, and Text

Virpi Oksman

INTRODUCTION

If a user had the possibility to watch the latest television news from the mobile phone, or listen to the news on the radio or read text news with the mobile, what would he or she choose? How would that differ with mobile entertainment content? In what cases would the user opt for mobile TV content and when would text or radio be chosen?

Mobile phones have been evolving into versatile multimedia devices that integrate different media forms, channels, and delivery systems. Yet it is not quite clear how users will perceive the functional blending of mobile communication—which people seem to experience more or less as a personal form of communication—and mass media broadcasting. In this light, this chapter examines users' mobile-media use and choices (text, audio, or video) in different everyday contexts. The data for the research is based on ongoing empirical research including field tests on mobile-media service prototypes in Finland in 2006 and 2007. The media service combined text, audio, and video and included both news and entertainment contents.

The empirical research shows that a device optimized for voice and text communication can offer users an interesting visual experience such as mobile TV news. It was interesting to note that the mobile-media contents could be used in various situations and times. Mobile TV use was an engaging activity: one or both hands were used to hold the mobile device, and users really concentrated on watching the small screen. Even at home , users did not leave the mobile TV on for "background noise"—the sound levels were considered very good, but users still held the phone in their hands and most often did not do anything else while watching. In that sense, audio and text news were seen as easier channels than video: only eyes or ears, not both of them, were needed.

Clearly, the mobile phone is evolving from a simple communication tool into a ubiquitous device for fulfilling various everyday life needs. Mobile phones can handle a variety of different media contents, and the mobile phone has become a medium used for various forms of personal, community, and mass communication. Mass communication content is increasingly available in personal mobile applications, which contributes to a blurring of the traditional boundary between personal and mass communication. The mobile phone has become truly "a new information medium" as well as a device to "harvest from ever-increasing palette of the digital domain."[1]

Of course, there are still many open questions related to the user experiences and user requirements of mobile television news and other media services. Users' preferences for the different uses of mobile media are being studied from various viewpoints. For example, mobile television has been studied by users' content choices,[2] viewing context and viewing times,[3] and video quality requirements.[4] It appears that the "killer application" for mobile media seems to be the opportunity to consume media at anytime, independently from the conventional broadcasting times.[5]

In our ongoing research project, we turned around the question of how interesting users find mobile television content in general and asked instead what media presentation type—text, audio, or video—users would choose in different situations and contexts, if they had the opportunity to make a choice. With this approach, we hope to gain new insight into users' requirements and expectations of different mobile-media contents and forms. We suggest that the combination of different media forms within the same mobile service is important because it offers users the ability to select the media format best suited to the situation at hand. In addition, in order to understand why people choose to employ or ignore certain mobile-media services and forms, it is meaningful to consider the role of mobile phones in users' everyday lives and activities.

THE MOBILE PHONE—A MEDIUM IN ITSELF

The role of the mobile phone as a personal mobile medium has been seen as a tightening of connections especially between close, already existing social networks; it makes possible to practice different kind of synchronous, more intense mediated social life than before with carefully selected social networks.[6] Japanese mobile communication researchers have called this phenomenon as "telecocooning," which refers to "a zone of intimacy in which people can maintain their relationships with others who they have already encountered without being restricted by geography and time"[7] and "the production of social identities though small, insular social groups."[8]

Moreover, as mobile communications are still relatively new technologies, they are often subject to reinterpretation. According to Janey Gordon, the mobile phone's role can be more far reaching than simple personal communications. Mobile phones played a pivotal role in world events such as in Manila 2001 and on 11 September in New York; highlighting their far-reaching effects on communication and culture.[9] The role of the mobile phone in public and also civic activities is central in Howard Rheingold's vision of the phenomenon, which he calls "smart mobs."[10] The notion of smart mobs means that people, through the facilitation of mobile technologies, can act in organized way to achieve some common goals. Rheingold observed that behavior, via mobile technologies, was affording different social networks abilities to increasingly engage and intervene in media cultures. For instance, with practices such as "citizen journalism," which refers to the phenomenon that ordinary citizens play an active role in reporting, collecting, and spreading news and information.[11] The importance of the mobile phone as a tool of citizen journalism is evident considering how its mass use can provide the spreading of news in situations and modes of delivery and distribution that the classic media cannot always compete.[12] However, the mobile phone is not just an empowering and liberating agent, for it can be used also as a means of social control and supervision. Thus, mobile technologies often perform in "Janus-faced" ways that give rise to multiple implications.[13]

Communication technologies are multipurpose tools that can change their major functions during time. Originally the landline telephone was used as a broadcasting medium since the 1880s, and it was not intended as a medium of bilateral communication. The important lesson from the history of the phone is that the users will finally determine how communication technologies are used, and often users considered as marginal may find successful uses—such as women and sociability and teenagers and SMS.[14]

The diversity of the mobile phone gives no clear function and its role in the changing society and social interaction depends on other developments.[15] The mobile phone as a medium does not just follow developments of other media, but it has become a medium in itself, strongly created by its different and changing uses. Recently, the mediatization process of the mobile phone has been powerful. According to Leopoldina Fortunati, the mediatization process can be defined as the impact of the classic mass to the fixed and mobile net. Especially the transition of GSM to the 3G has been seen as signaling the evolution of the mediatization to the mobile phone, the culmination of which is the ability to follow TV broadcasts on the mobile.[16] Digital media convergence has enriched the mobile phone with various media functions, forms, channels, and delivery methods. There is hardly any other single device which has converged so many different technologies and functions.[17] In addition, besides the impact of the classic media to the mobile phone, the mobile phone has some special characters as a medium, as it is strongly connected to the user-generated content production. Text and image messaging and videophone features have generated new kinds of interpersonal communication forms and cultures in the daily lives of people. Lately, these communication forms have started to fuse with classic media—such as the videos taken by mobile-phone cameras—to form a significant part of some media companies' Web portals. Moreover, there is also *mobilization* of conventional TV—the mobile-phone-made SMS messages are given a special role as a part of current affairs TV programs as they appear as the public's comments on the TV screen. Besides this, there is also "mobilization" of the press; in some national newspapers in Finland there are columns which consist purely of readers' opinions and comments sent by SMS.

The degrees of mutual influence between different media technologies is inevitable, both because the communicative environment squeezes them into one space and because people make use of various means of communication in everyday life.[18] Traditional media have set the standards for information quality. It appears that mobile-media users also expect accurate, timely, and high-quality information delivered through reliable channels. However, if a media channel proves to be inaccurate or technically too fragile, the consumer will soon abandon it for a better alternative.[19]

Taking this into consideration, this chapter will apply an integration approach to the use of mobile phone as a "postbroadcast" medium. According to David Holmes, the way in which individuals find connection with different media forms can be shown to be interdependent. For instance, network media become meaningful because of broadcast, and broadcast becomes meaningful in the context of network media.[20] This means here that the role of mobile-media content is also discussed in relation to other media usage. Different media forms are discussed as discursive interstitials, rather than as distinct fields in the users' everyday lives.

THE RESEARCH QUESTIONS

The core question of mobile-media consumption lies in user appropriation: for what purposes do the users wish to use the mobile news and TV feature in the context of daily life? According to Eija Kaasinen,[21] key factors that affect users' willingness to start using any particular mobile service are the value of the service to the user, its *ease of use,* user's *trust* towards the service (provider), and its *ease of adoption.* The value of the service refers to its (immediate) usefulness, but it can also be understood as referring to more abstract value, for example, social prestige or a sense of shared experience. Giving the importance of the user appropriation of the mobile services, the following research questions will be explored here:

How do people want to use different media forms in their mobile phones?

How does the mobile-media use relate to other media use in the daily life?

ABOUT THE RESEARCH

The purpose of the first field test was to find out what kinds of mobile content (video, audio, and text) people choose in different situations. A media service prototype was developed for this purpose. The first prototype contained a simple browser and on-demand news content recycled from television, radio, and teletext services (produced by YLE, Finnish Broadcasting Company).

Figure 10.1   Podrace media service prototype combines text, audio, and video.

In the field study, qualitative and quantitative methods were combined to make sure that adequate data are collected. Semistructured interviews and media diaries helped us understand users' media habits and how they voice their expectations and preferences. Log data reveal the time and duration of actual occurrences of service use. We also asked the users to take some photos with the camera phone about the situations in which they might use the mobile news service, and of other places, things, or contexts that are important for them. This helps us gain an understanding about the role of media and technology in the users' everyday lives.

The first field test started in March 2006 with ten informants who used the service with 3G phones (Nokia 6630 and N70) for one month. Before the test period, the users were interviewed and they received information concerning the test. Demographic data and media user profiles were gathered from the informants. The ages of the informants ranged from twenty-three to fifty-six. All of them worked at least part time and had used mobile services before. During the test period, the informants reported their user experiences in a test diary. After the test, users were asked to fill in a usability evaluation form and they were interviewed again.

Similar kinds of methods were used in the second field test, which started in October 2006 and ended in December, with the exception that the log data were not possible to collect from these applications. The purpose of the second test was to test new mobile services with more varied contents and also to evaluate the quality differences between 3G TV and DVB-H TV. Ten families tested the services with Nokia N92 for a period of one month. The tested mobile TV services consisted of a wide range of different kinds of contents: from main TV channels to sports news channels and from fashion TV to user generated contents.

Both test groups consisted of persons who have been using mobile phones and mobile services actively during the last few years. They were also keen news and media followers, but each had different kinds of media user profiles. Some of them were very loyal newspaper readers while others regarded the Internet or TV as the best news or entertainment source. Testers had different kinds of hobbies, lifestyles, and interests. During the test, they carried the testing phone as their primary mobile, using it for both professional and personal communication.

THE CONTEXTS OF USE

According to the mobile news service log data gathered in the first field study, the mobile-media service was used several times daily during the research period. Out of the various media formats on offer, television was most widely used. When the users had a possibility to choose to receive news in different media forms in their mobiles, text format was the most often opened option. Typically, users perceived text news format as the most convenient for various kinds of situations and especially fitted for quick news headlines updates "on the go." Text-based news format was also discovered as less vulnerable to the functioning problems of the 3G network. However, regarding the total amount of time, video news format was used longest. It seems that video news viewing is actually done more rarely than opening text news, but at once, when the reception is good; people would like to see the news broadcasts a bit longer than just at a glance.

Situations where the user is unattainable by regular media were considered as best suited for the use of the mobile news service. For example, while spending a holiday at a location where no fresh newspaper was available, an assiduous news follower began to crave an additional link to the outside world. According to test diaries and interviews, the service was used in different everyday contexts: in the bus on the way to work, on a train, on vacations, coffee breaks at work. and at home before bed. Mobile TV obviously possesses some novelty value that was present in the usage situations.

> The thing I used the most was TV news during my coffee break at work. That way I came to show my friends that I had a TV in my mobile. Some thought, well, you always have to have the latest gadget. Others said, ok, that's interesting, but the screen is very small. You could make out what it is, but the size of the screen caused a little doubt whether or not it's worth it. (Man, 56)

The mobile TV service was mainly used by owners as an individualized and personal media form. However, while watching TV, the device became occasionally a more social tool as a point of novelty. After this phrase, it was mostly used by one person only. One test user decided to demonstrate the mobile television while coming from the ice hockey game by train.

The train was quite packed by people coming from the match. Then I turned on the sport news, and everyone got quiet, curious to see how the game was reported by the mobile news. (Man, 52)

The use of mobile TV raised discussion in the immediate circle of the testers. The general view was that its use was best suited for situations where other media were not available or where people found themselves with extra time on their hands. Users described it as a nice way of passing the time. Users mentioned that in noisy environments, like in a bus or in traffic, media types other than TV might be easier to use. The use of earpieces in a public place was seen as less awkward than it would have been before; however, mobile TV would not be used without earpieces, for example, in a bus, and sometimes it would be an extra effort to find and attach them to the phone.

The ability to select the media format most appropriate for the situation at hand was considered important. Audio was perceived as suitable for situations where the user was mobile him/herself, for example, while walking, cycling, or roller skating. For situations when the user was sitting or standing still, the media form selected was more likely to be illustrated news, text, or video.

Figure 10.2 The mobile prime place. The mobile TV was used often when moving from one place to another. Source: Author.

THE MOST INTERESTING MEDIA CONTENTS IN MOBILES

Clearly, as the earlier studies also show, news was considered as one of the most interesting media contents on mobiles.[22] The categories of domestic (25 percent), sports (15 percent), and foreign news (9 percent) attracted the most interest. In addition, TV program guides were checked quite often (11 percent). Local news and children's sections were read more randomly. There was high demand only for the latest news—the older news from archive was barely read or watched at all. Regarding the entertainment contents, mostly same contents were watched as on regular TV, but also special channels were liked, for example, a channel focusing on local cultural events.

In the second field test, video-watching durations were considerably longer than on news field trial, approximately from five to ten minutes (according to the testers' own report). Besides the new contents, also 3G and DVB-H reception difference may have an influence on session durations. The test users saw a clear difference in image, sound, and text quality between 3G TV and DVB-H. In all things, the DVB-H quality was seen as superior. The DVB-H network functioned more securely and it covered more areas—and thus more often available for use in different places. In general, because of the 3G network problems the service worked sometimes quite unsteadily and thus mobile TV watching was not always possible when wanted. Also, weather was suspected to have influence on mobile TV reception.

> Yesterday, when it was raining heavily, I was not able to watch 3G TV on the bus, as the reception kept cutting all the time . . . So perhaps it is possible that the weather affects data transfer? (Man, 56)

Also, certain reality TV series that the users want to keep up with real time were mentioned as interesting contents. During the test period an obvious example of this was the reality TV show *Big Brother*, which, in addition to television, was followed up through other media. However, the mobile TV was also expected to offer something more than conventional TV. Some test users found interesting the interactive mobile TV services, like buying tickets. Most likely the successful mobile-media services will combine both user-generated and professionally generated features.

> It was surprisingly interesting. We love to go to the theatre often, and I could eagerly buy tickets through this service. It would also be nice, if you could create your own profile, and you would receive exactly the services which interest you. (Man, 36)

MOBILE PRIME TIME

The average viewing time for mobile TV news video was quite short (a median for video duration was one minute forty-three seconds). The

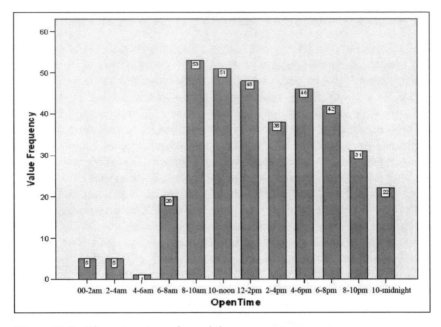

Figure 10.3   The prime time of a mobile news service.

mobile-media-service use spread relatively evenly for the whole day, although use was more frequent during the mornings (from 8 to 10) and before the noon (from 10 to 12) and early in the evenings (from 4 to 6).

Users appreciated the ability to watch news any time they liked to, without strictly scheduled agenda. "It is very handy indeed that you are able to watch the news whenever it is suitable for you. You don't have to care about times" (Woman, 43).

The service use was most active in the beginning of the week. The usage tended to become less active during the weekends—it was remarkably low on Sundays, probably because people were at home and had the possibility to watch regular TV broadcasts.

RELIABILITY OF DIFFERENT MEDIA

The mobile news service was considered relatively fast in comparison with other media. The test group compared and assessed the reliability of different media through a variety of viewpoints. Some considered that despite its reliability the newspaper may be unable to compete in speed with electronic media occasionally; printed news may already be dated as it is released. Some continued to value the reliability of newspapers despite their slowness: electronic media may publish anything with no verification.

The image of newspapers is reliable but it is incredibly dated. It's pretty slow compared to the Internet in that the situation can have changed by the time the paper comes out. (Woman, 26)

These days it is possible for anyone to produce electronic media. If we talk about reliable media, that for me is print media, national newspapers (such as *Helsingin Sanomat*), or the major TV channels. If you get it through some other channel, you start thinking is this a joke or something somebody's made up. (Man, 42)

Thus, the different media are mutually constitutive in the users' everyday lives. The traditional print media are meaningful because they differ from electronic media. Newspapers are considered quite slow compared to the electronic media, but they are considered as the most reliable media. Crucially, text was generally considered the most reliable form of media also in mobiles. It is suited for many different situations and was accessible in situations where other forms of media are not, for instance, due to network problems. It appears that the traditional media were also used as "the official evidence" that the news heard or read during the day from the mobile phone and the Internet was correct.

It also appeared that watching broadcast on the small screen was quite demanding physically. Mobile TV use was experienced as an engaging activity: often one or both hands were used to hold the mobile device, and users really concentrated on watching the small screen. One test user described the watching of mobile news as a demanding job in his media diary: "My hands are starting to get really tired after watching long news as I have to hold the phone in my hands" (Man, 24).

Even at home, users did not leave the mobile TV on for "background noise"—the sound levels were considered very good, but users still held the phone in their hands and most often did not do anything else while watching. In that sense, audio and text news were seen as easier channels than video: only eyes or ears, not both of them, were needed. The test users were worried about becoming absorbed in mobile multimedia content, which requires their visual attention, so the mobile services were not used while driving or in other more demanding activities in public spaces.

Thus mobile-media services were seen as best suited for quick updates of news and reality series and the viewing of trailers or advertisements for upcoming TV shows: not many people would be interested in watching a whole episode of a TV show or a movie on the screen of a mobile at this stage.

CONCLUSION

From the empirical case study we can see that a device optimized for voice and text communication can offer users an interesting visual experience

such as fresh TV news and entertaining contents. All the media types—
text, audio, and video—proved to be useful in mobiles. Especially the wide
interest in mobile TV could be perhaps seen as part of a broader tendency
towards "personalization" in the media environment; nowadays people
would like to see the news when it is most suitable for their own schedules,
and some family members may even wish to have their own small *personal*
mobile-media service at home besides their regular TV sets.

The mobile phone as media seems to be suited for many different situations. Mobility (which here refers to functional dimensions of portability and freedom from social and geographical constraints),[23] diversity, and
real-time effect are considered as the most important characteristics of the
service. The combination of them makes the use of the mobile news service
different from any other media use. Users appreciated updated information and information-rich media forms for mobile news delivery. There was
high demand only for the latest news in mobiles. The real-time effect was
considered as important. Users also appreciated fast functions and easy
usability. Compared to the situation in earlier studies on mobile video
content,[24] the use of earpieces with the mobile phone has become increasingly common and makes it more convenient to follow media contents (for
instance, in public transport) without disturbing others.

The mobile TV service was mainly used by the owner of the device as
an individualized and personal media form. As personal communication
devices are turning into multimedia communication devices delivering
news and other mass media contents, new questions about user experience
challenges will emerge. Two users interestingly identified that they would
expect the user interface to display a new functional or visual idea. They
had recently started using a text-based syndicated news browser that has
its own "smooth" scroll implementation, which they had enjoyed using.
For these users, the new scrolling implementation signified the service
providers' investment and commitment to developing a good service, and
this increased users' positive attitude towards the service. Users appreciated condensed information and media forms for mobile TV and news
delivery. Most users looked the headlines or followed news several times
a day—much more often than the traditional TV and news prime times
would allow.

The test group of users considered some technical problems as causing
the major problems in the mobile media and especially in the mobile TV
use. Thus, usability issues regarding the small-screen user interfaces will be
particularly central and the good quality of reception is crucial. In the long
run, it will also be crucial to discover what kind of existing and new media
formats and distribution channels will best suit mobile media. Regarding
the entertainment services, mobile video is at the moment best suited for
quick updates of TV shows and perhaps for some "minisodes."[25] Especially
certain reality TV series that the users want to keep up with in real time
were mentioned as interesting contents.

NOTES

1. Harvey May and Greg Hearn, "The Mobile Phone as Media," *International Journal of Cultural Studies*, 8(2) (2005): 200.
2. Caj Södergård, ed., *Mobile Television—Technology and User Experiences*, Report on the Mobile-TV Project (Helsinki: VTT, Edita Prima Oy, 2002); Hendrik Knoche and John McCarthy, "Good News for Mobile TV," paper delivered to *World Wireless Research Forum 14*, San Diego, CA, July 7–8, 2005; Virpi Oksman, Elina Noppari, Antti Tammela, Maarit Mäkinen, and Ville Ollikainen, "Mobile TV in Everyday Life Contexts: Individual Entertainment or Shared Experiences?" in Pablo Caesar, Konstantinos Chorianopoulos, and Jens F. Jensen, eds., *Interactive TV: A Shared Experience*, proceedings of the 5th Euro ITV Conference (Amsterdam, 2007).
3. Birgit Plodere and Marianne Obrist, "Towards Integrated Acceptance Model for the Design and Reflection of Interactive TV Research," *Proceedings of the 4th EuroITV Conference*, Athens, 2006; Juri Mäki, *Finnish Mobile TV Pilot: Results* (Helsinki: Research International Finland, 2005); http://www.mobiletv.nokia.com/pilots/finland/files/RI_Press.pdf Virpi Oksman et al., "Mobile TV in Everyday Life Contexts."
4. Henrik Knoche and John McCarthy, "Design Requirements for Mobile TV," ACM International Conference Proceedings Series of *The 7th International Conference on Human Computer Interaction with Mobile*; Plodere and Obrist, "Towards Integrated Acceptance Model," Juri Mäki, *Finnish Mobile TV Pilot*; Virpi Oksman et al., *Mobile TV in Everyday Life Contexts*; Satu Jumisko-Pyykkö and Jukka Häkkinen, " 'I would like to see the face and at least hear the voice': Effects of Screen Size and Audio-Video Bitrate Ratio on Perception of Quality in Mobile Television," *Proceedings of the 4th EuroITV Conference* (Athens, 2006).
5. Södergård, *Mobile Television*.
6. Hans Geser, "Towards a Sociology of the Mobile Phone," http://socio.ch/mobile/t_gescr1.htm.
7. Ichiyo Habuchi, "Accelerating Reflexivity," in *Personal, Portable, Pedestrian: Mobile Phones in Japanese Life*, ed. Mizuko Ito, Daisuke Okabe, and Misa Matsuda (Cambridge, MA, and London: MIT Press, 2005), 167.
8. Mizuko Ito, "Introduction: Personal, Portable, Pedestrian", in *Personal, Portable, Pedestrian: Mobile Phones in Japanese Life*, ed. Mizuko Ito, Daisuke Okabe, and Misa Matsuda (Cambridge, MA, and London: MIT Press, 2005), 10.
9. Janey Gordon, "The Mobile Phone: An Artefact of Popular Culture and a Tool for Public Sphere," *Convergence*, 8 (2002): 15.
10. Howard Rheingold, *Smart Mobs: The Next Social Revolution* (Cambridge, MA: Perseus Publishing, 2002).
11. Jens F. Jensen, "User Generated Content—A Mega-trend in the New Media Landscape," a presentation at *EURO ITV 2007 conference*, Amsterdam, the Netherlands, 2007.
12. Leopoldina Fortunati, "The Medialization of the Net and Internetization of the Mass Media," *International Communication Gazette*, 67(27) (2005): 35.
13. Michael Arnold, "On the Phenomenology of Technology: The 'Janus Faces' of Mobile Phones," *Information and Organization*, 13 (2003): 234.
14. Amparo Lasen, *The Social Shaping of Fixed and Mobile Networks: A Historical Comparison* (DWRC, University of Surrey, 2002, http://www.dwrc.surrey.ac.uk/Portals/0/HistComp.pdf).

15. Sadie Plant, "On the Mobile," International Telecommunications Union's Background paper *Social and Human Considerations for a More Mobile World*, 2004.
16. Fortunati, "Medialization."
17. Gordon, "The Mobile Phone," 8; Fortunati, "Medialization," 27.
18. Fortunati, "Medialization," 28.
19. Andreas Nilsson, Urban Nuldén, and Daniel Olsson, "Mobile Media, The Convergence of Media and Mobile Communications," *Convergence*, 7(1) (2001): 38.
20. David Holmes, *Communication Theory: Media, Technology and Society* (London: Sage, 2005).
21. Eija Kaasinen, *User Acceptance of Mobile Services—Value, Ease of Use, Trust and Ease of Adoption* (Helsinki: VTT Information Technology, 2005), 70.
22. Södergard, *Mobile Television.*
23. Kenichi Fujimoto, "The Third-Stage Paradigm: Territory Machines from the Girls' Pager Revolution to Mobile Aesthetics," in *Personal, Portable, Pedestrian: Mobile Phones in Japanese Life*, ed. Mizuko Ito, Daisuke Okabe, and Misa Matsuda (Cambridge, MA, and London: MIT Press, 2005), 80.
24. Petteri Repo, Kaarina Hyvönen, Mika Pantzar, and Päivi Timonen, *Mobiili video* [Mobile video] (Helsinki: Kuluttajatutkimuskeskus, julkaisuja, 2003).
25. Bill Carter, "Coming Online Soon: The Five-Minute 'Charlie's Angels,' " *The New York Times*, April 30, 2007, http://www.nytimes.com/2007/04/30/technology/30sony.html?ex=1335585600&en=a0aa68272a6e520a&ei=5090&partner=rssuserland&emc=rss.

11 New Economics for the New Media

Stuart Cunningham and Jason Potts

New, mobile and Internet media have attracted significant attention not only from cultural and media studies but also from new fields of economics and business analysis concerned with searching out the conditions for "creative disruption" and the role of "disruptive technologies" in contemporary societies. This chapter introduces these new strands of economic thinking into the debate about the impacts of new, mobile, and Internet media, while also exploring the implications of this approach for positioning the creative industries as a driver of innovation. It does this by offering four models for conceptualizing the role of creative industries in influencing economic growth and change.

New media—and mobile media because they particularly provide affordance for social networks—are crucial in this conceptualization. We characterize them as integrating and transforming new *technologies* into new *services* and introducing variety in the economy because of continuous flows of novelty (in content or design, for example). This gives rise to the proposition that creative industries (and new media in particular) can be freshly defined in terms of "social network markets."

Over the last ten years a new, "evolutionary" approach to economic analysis has become prominent. It focuses on the dynamics of the economic system under conditions of variety generation, enterprise competition, and selection and self-organization.[1] Most of the empirical and theoretical work so far undertaken in modern evolutionary economics has focused on manufacturing and high-technology sectors, as have most analyses of the sources of innovation in contemporary economies. There is little yet that seeks to apply this new framework to the economic analysis of the creative industries (CIs), although some of the most interesting is focused on the broader question of innovation in services.[2] The core advance that this approach might facilitate is to understand creative industries as an emergent, innovative part of the services sector of the economy, rather than presenting them as an exception to mainstream industries, as "not just another business." The professional interest group Focus on Creative Industries (FOCI) in the United Kingdom captured what this shift might imply well:

> Whilst FOCI welcomes the recognition of the strong economic contribution made by the creative industries in terms of wealth creation and employment, we would also keenly stress that this sector is very different from traditional industries. They deal in value and values, signs and symbols; they are multi-skilled and fluid; they move between niches and create hybrids; they are multi-national and they thrive on the margins of economic activity; they mix up making money and making meaning. The challenge of the creative industries is the challenge of a new form of economic understanding—they are not "catching up" with serious, mainstream industries, they are setting the templates which these industries will follow.[3]

Evolutionary economics focuses on the ways economies *grow*, as complex open systems, rather than by optimizing allocative efficiencies. It is also offers a clearer understanding of the way in which new technologies are integrated into an economy and the restructuring of organizations, industries, markets, communities, and lifestyles the evolutionary growth process of economies requires. The creative industries' complex contractual and organizational structures, inherent uncertainty, power-law revenue streams, and high rates of experimentation—like Richard Caves's "basic economic properties of creative activities"[4]—suggest that they may be "pure" cases of service-based competitive enterprise in an open, uncertain environment. It is the purpose of this chapter to outline how this case might be developed in the light of possible models of the relation of the creative industries to the wider economy, and where new, mobile, and Internet media might fit.

"Creative industries" is a relatively new analytic definition of the industrial components of the economy in which creativity is an input and content or intellectual property is the output. Policy attention has focused on the size and growth of this sector. The creative industries have thus come to be newly represented as a significant and rapidly growing set of industries; an important sector, in other words, for policy attention. Globally, mapping exercises have concurred that the creative industries are indeed "economically significant" and legitimately comparable to other high-profile sectors in terms of their contribution to income, employment, and trade. The CIs, by implication, are argued to deserve policy attention (and support) in proportion to that measured significance.

But the value of these exercises can be limited because of the standard economic approach to industry analysis, which is based on the assumption of "general equilibrium." Measuring a static slice in time doesn't tell us why and how change is occurring. It is a matter of *political* expediency to afford an industrial sector policy attention and support in proportion to the share of income (or jobs, or foreign exchange) it generates, not a matter of economic logic.

Instead, we pose a different question: what is the role of the CIs in the process of driving and facilitating change, as evidenced by its dynamic parameters and degree of embedding in the broader economy? There is an emerging suspicion that the dynamic significance of the creative industries may well be of greater analytic and policy significance than its static role with respect to the level of jobs, aggregate output, exports, and cultural value.

We seek to pose this question directly: what is the dynamic relation between the creative industries and the rest of the economy? This means how a change in one effects the other. The four models proposed here are the four possible answers to this question: namely (1) negative, (2) neutral, (3) positive and (4) emergent. Each of these hypotheses or analytic models then parlays into a very different policy model: in the first case a welfare subsidy is required; in the second, standard industry policy; in the third, investment and growth policy; and in the fourth, innovation policy is best.

MODEL 1: *THE WELFARE MODEL (SPECIAL INDUSTRY)*

In model 1, the creative industries (although specifically the cultural industries, which are a subset of the CIs) are hypothesized to have a net negative impact on the economy, such that they consume more resources than they produce. To the extent that they continue to exist in a market economy, their value must lie at least in part beyond market value. In model 1, the cultural industries are essentially a "merit good" sector that produces commodities that are welfare enhancing, but are only economically viable with a transfer of resources from the rest of the economy.

This is typical of what are called 'public goods' and for which the economic justification for allocative restitution therefore ultimately rests on a *market failure* argument. Policy is then calibrated to estimates of this nonmarket value. If model 1 is true, then policy prescriptions should rightfully center about income and resource reallocation, market or regulatory intervention, and so on, in order to protect an inherently valuable asset (i.e., cultural production) that is naturally and continually under threat in a market economy.

It is broadly accepted by scholars of cultural economics[5] and supported by a raft of nonmarket valuation studies[6] that the overall contribution of cultural goods and services of these kinds is, on the whole, mostly positive. This is an unsurprising and indeed edifying result that accords with intuition. However, in economic terms, evidence for model 1 is built from: high levels and rates of negative profit among firms; low total factor productivity; persistently lower income to factors of production in the CIs as compared to other industries; or other indications that the survival of organizations within the CIs is critically dependent upon resource transfers from the rest of the economy to maintain prices or demand. If model 1 is true, we would expect to observe not just an economically stagnant or low-growth sector but one with lower performance levels (e.g., return on investment, incomes, etc.). It is usually accepted that this model fits most accurately the arts end of the creative industries spectrum and that cultural economics has largely been developed to address issues arising from these assumptions. These observations, however, do not fit many other parts of the CI continuum.

MODEL 2: THE COMPETITIVE MODEL (JUST ANOTHER INDUSTRY)

Model 2 differs from model 1 in presuming that the creative industries are not economic laggards, nor providers of special goods of higher significance, but are effectively "just another industry." It presumes that the growth impact is neutral, such that CIs contribute in aggregate no more or less technological change (i.e., origination of new ideas but also their adoption and retention) than the average of other sectors. This model does not argue that the CIs have no effect on income or productivity growth, as that is trivially false, but that their effect is on *par* with all other sectors. If the CIs were at least as dynamic as other industries, then we should expect model 2 to be true. This model might be seen to fit best the established media industry sectors that are mature, experiencing static growth, or are in relative decline (some parts of publishing and print, broadcasting, and perhaps the commercial end of film, with the subsidized end of film fitting model 1).

If so, this would also imply that there are no economic welfare gains to be had by special policy treatment. This implicitly supposes that cultural/creative goods are "normal goods" that rational consumers would substitute between those from other sectors as they vary in price in order to equalize marginal utility. In this case, an expansion of the CI sector would have no aggregate welfare benefit distinct from expansion of any other sector.

Model 2 does not exclude the possibility that the economics of the creative industries are different and distinct. Richard Caves and Art de Vany have talked of extreme levels of demand uncertainty, power-law revenue models, tendencies toward monopoly, complex labor markets and property rights, endemic holdup problems, information asymmetries, highly strategic factor markets, and so on.[7] Rather, it emphasizes that these "coordination problems" will be eventually be solved under competitive conditions, just as the special circumstances of other industries lead them to discover specific institutional arrangements and coordination structures. Model 2 thus thinks that, ultimately, these special features are no different to the special problems of most other industries, such as energy or tourism, which also have interesting features associated with scale, coordination, uncertainty, networks, and so forth. The competitive model thus hypothesizes that the CIs will have comparable industry performance statistics to other sectors.[8]

The CIs then require no special policy treatment, just the consistent application of policy mechanisms extended to other industries. If model 2 is true, then the CI policy focus should not be about targeted resource reallocation or special intervention but rather with the plea for consistent industrial treatment. Evidence of the competitive model would come from the equivalence of CI economic indicators with those of the whole economy through evidence of normal competition and enterprise.

MODEL 3: *THE GROWTH MODEL (CIS AS INPUTS INTO THE BROADER ECONOMY)*

Model 3 explicitly proposes a positive economic relation between growth in the CIs and growth in the aggregate economy. In this model the CIs are one of the growth "drivers" of contemporary economies, in a similar way agriculture was in the early twentieth century, manufacturing was in the 1950s–60s, and information and communication technologies (ICT) was through the 1980s–90s. The possible reasons include the notion that the CIs introduce novel ideas into the economy that percolate to other sectors (e.g., new uses, new designs), or the CIs facilitate the adoption and retention of new ideas and technologies in other sectors (e.g., ICT). The key difference from models 1 and 2 is that in model 3 the CIs are causally involved in the *growth* of the economy. The supply-side interpretation of this model emphasizes that the CIs carry new ideas to the aggregate economy. The demand-side interpretation emphasizes how growth in the aggregate economy causes a proportionate (or even disproportionate) increase in demand for CI services.

In both supply-side and demand-side cases, policy should properly treat the CIs as a "special sector." This is not because it is economically significant in itself but because it powers the growth of other sectors. This may plausibly lead to intervention, but unlike model 1, which also affords special treatment but for subsidy reasons, the ostensible purpose is to *invest* in economic growth and the development of capacity to meet growth in demand. If model 3 is true, a clear economic case exists for redirecting resources and not just for the benefit of the CIs per se but for wider benefit. The creative industries, in this view, are a winner to be backed.

Evidence for this model would accrue from association of the CIs with growth not just in jobs and commodities (as in model 2) but also in *new* types of jobs and *new* sorts of commodities and services. Model 3 proposes the CIs as growth drivers not because of operational expenditure multipliers but via the introduction and development of new technologies and new uses for, and methods of absorbing and diffusing of, technologies. The CIs would be thus assumed to create new industries and market niches that could stabilize and develop extant industries. This is the opposite of model 1, in which economic growth suffers (if only marginally) when there is such continued investment. Model 3 argues that the CIs are good for the economy because they introduce and process new ideas that drive economic growth.

This model accommodates design as an input factor into the economy, interactive leisure software like games, and also mobile and Internet media. These exemplify input impact, such as games providing models for new generation education and learning paradigms or for simulation and virtual reality training in defense. It is evidenced by the positive correlation between design intensity in firms and their stock market performance.[9] It also is suggested by the growing proportion of creative occupations "embedded" in the broader economy. But, as we shall see, it is perhaps best exemplified by the huge growth of mobile and Internet media use and content creation and the unexpected (on the supply side) uses to which such activity and inventiveness has been put.

MODEL 4: *THE INNOVATION MODEL*

Although these three models might seem exhaustive of analytic possibilities, a fourth model is also possible as an emergent dimension. Rather than thinking of the CIs as an economic subset "driving" growth in the whole economy, as in model 3, the CIs may not be well characterized as a sector per se but rather as an element of the *innovation system* of the whole economy. The value of the CIs, in this view, is not in terms of their aggregate contribution to the economy either sectorally or as a whole (as in models 1–3) but due to their contribution to the coordination of new ideas or technologies and thus to the process of change. In this view, the CIs are misspecified as an industry in the first place but better modeled as a complex system that derives its "economic value" from the facilitation of change and growth; a system that manufactures attention, complexity, identity, and adaptation though the primary resource of creativity.

Similar to model 1, model 4 ventures an element of special pleading akin to that of science, education, and technology in the *national systems of innovation* approach[10] in which the CIs originate and coordinate change in the knowledge base of the economy. Culture and creative action is not an industry per se, in this view, but rather a mechanism of all industries. Like technical education and science, the CIs are viewed as an essential innovative component of a modern postindustrial economy and a primary source of competitive advantage. Change in the CIs therefore produces structural and not just operational change in the economy. It follows that new opportunities and possibilities will thus emerge from the CIs for which the economic and welfare effects cannot be known in advance. According to model 4, the CIs do not drive economic growth directly, as might a boom in the primary resource sector or the housing market, but rather indirectly via the facilitation of change in the economic order. For example, it might be possible to define the CIs not as an industry sector but as a space of economic activity in which actors, markets, and organizations are predominantly shaped by social networks and thus are the kind of "switching yard" (to use a transport metaphor) between social and commercial activity.

If model 4 is true, the CIs should be understood as a significant component of *innovation systems* that drive and coordinate the growth of knowledge processes underpinning economic evolution.[11] This suggests innovation policy is a superior instrument to competition or industry policy, as well as aligning CI policy with science and education policy. If model 4 has validity, it justifies an "elitist" aspect to CI policy in the same way that old-school versions of cultural policy justified the development of culture as a public good. But unlike the cultural value of museums or classical arts, which seeks value through the maintenance of past knowledge, CI value lies in the development and adoption of new knowledge, and so is focused on experimentation and difference rather than conservation and equality.

WHAT ROLE FOR NEW, MOBILE, AND INTERNET MEDIA?

The relative growth and "embedding" of the CIs, giving credence to the notion of a "creative" economy, is not an anomaly but what open-system economic theory predicts based on the effects of technological change (i.e., endogenous growth) and a changed consumption set consistent with increased income.[12] This manner of growth in the CIs is predicated on: rising affluence, shifting aggregate expenditure toward the CIs as part of general growth and consumption of services; the related rise in levels of human capital (higher levels of education, etc.), which permits greater specialization; the phenomenal growth in ICT, which is the technology base of the CIs; access to global markets both in demand and factor mobility; and postindustrial and media-cultural identity reasons. The drivers of economic growth are multiple and interacting (or coevolving); the CIs are a necessary but not sufficient feature of contemporary economic growth.

The uses to which new, mobile, and Internet media are being put play a crucial role in the contemporary adoption, absorption, and retention of new technologies, in challenging and changing the business models of many industries (not just the media industries), and in creating the conditions for what we are calling "social network markets."

Mobile media have been at the forefront of design-led competitiveness in new economy business models. Design has become a fundamental input into most products and services in the digital economy; it is one of the key instances of creativity-as-an-enabler in both the manufacturing and service sectors. Design brings a demand-driven focus and customer research, branding, and distribution to the fore. For high-growth creative and service industries, innovation is as much a user-driven as technology-driven process. The World Economic Forum's Global Competitiveness Reports have consistently shown that there is a distinct correlation between design intensity in enterprise activity and product development and broad economic competitiveness. The UK Design Council has produced data to show a correlation between design intensity and stock market performance over time across the general UK economy. Some of the headline examples of design intensity are mobile companies: Nokia for Finland, Phillips for The Netherlands, and Sony Ericsson for Japan/Sweden, Samsung for Korea, and so on.

In the telecommunications market, much growth occurs at the retail end through the huge and sometimes surprising uptake of new capacity and applications. Whereas a great deal of business practice remains production-centric—with the matter of design inputs often relegated to marketing and commercialization strategies—the world players in phone market regard design-for-branding as their core business. Nokia's move in the late 1990s from "the dominance of technical issues and the image of a sophisticated, and thus 'demanding,' technology" to a "design house for mobile communication" is exemplary.[13]

But mobile media models not only consolidate new economy business practice; they also challenge existing practice, in some ways quite fundamentally. This suggests new, mobile, and Internet media occupy important roles in both models 3 and 4. The explosion of user-originated content and user-led innovation has been platformed uniquely by new, mobile, and Internet media. And it's not necessarily the size of this sector—games are bigger than film; there is more user-generated content on the Internet than professionally produced and corporate content, and so on—though that indicates something significant. It's that they provide a classic instance of "creative destruction," upsetting the business models of the established communications conglomerates, introducing novelty into the system, and leading even Rupert Murdoch himself to presage the end of the days of the media mogul.

What might be specific about mobile media in this context? As a social technology, it is possible to note that mobile media are more broadly embedded across global region and demographic grouping than other new media technologies. In many regions of the world, mobile communication substitutes for fixed-line telecommunications and in most mobile is challenging fixed line for predominance. Many are prepared to predict that mobile may become the technology around which further convergence occurs. There is a closer integration of personal media (Short Message Service [SMS], Multimedia Messaging Service [MMS]) with social network formation in everyday life and thus the potential for a more dynamic coevolution of the relationship between the technology and social networks and emergent market formations based on them.

A good deal of the thinking around the challenges that user-originated content and user-led innovation provoke in contemporary business models, intellectual property regimes, and the workings of social networks relies on how new, mobile, and Internet media are operating. Chris Anderson[14] and

Mark Pesce[15] exposit the limitations of the mass market, blockbuster mentality that can be addressed by Internet-based harvesting of the "long tail" and exploitation of "hyperdistribution." Henry Jenkins points to the collective intelligence of social networks that drives the intense value adding of new media such as online games and the global critical social spaces that have come into being around mash-ups.[16] Charles Leadbeater's *We-Think: The Power of Mass Creativity*[17] explores diverse domains where the power of socially networked collective creation and communication is at work. The implications of this for legal and economic regimes is explored in rigorous detail by Yochai Benkler in *The Wealth of Networks*.[18] In the light of these developments, we are developing a new definition for the creative industries as "social network markets."[19]

The central fact about creative industries markets is that complex social networks play a much more significant coordination role than price signals. For evolutionary and complexity economists, who have long appreciated the economics of open-system processes as different from closed systems, this is unsurprising. What is new, however, is the suggestion that this might also apply not just to science and technology, which is the conventional basis of evolutionary economics, but also to the arts and culture. Markets for novelty as social networks are thus moved closer to the center of the economic analysis of innovation and growth. New, mobile, and Internet media are the principal means by which such social network markets operate.

The very act of consumer choice in creative industries is governed not just by the set of incentives described by conventional consumer demand theory but by the *choices of others*. Examples are given by Arthur,[20] de Vany and Walls,[21] Ormerod,[22] Kretschmer et al.,[23] and Beck.[24] An individual's payoff is an explicit function of the actions of others. Schelling described this entire set of issues as being "binary decisions with externalities."[25] There is overwhelming evidence[26] that this applies generally to the creative industries. Our new definition of the CIs therefore proceeds not in terms of individual "artistic" or creative novelty but rather in terms of many individuals' choices in the context of complex social networks of other individuals' choices. The CIs, then, are properly defined in terms of a class of economic choice theory in which the predominant fact is that, because of inherent novelty and uncertainty, decisions to both produce and consume are largely determined by the choice of others in a social network.

These social networks can function as markets, markets for ideas, for new uses for technologies, for social value creation and distinction.[27] So recognized, it becomes equally apparent that the CIs are also a crucible of new or emergent markets that, typically, arise from nonmarket dynamics (e.g., Internet affordances) and that often then stay at the complex borderland between social networks and established markets. For example, YouTube's social networks, which were then bought by Google and thus market conditions were brought to bear; MySpace is a similar example, which was recently bought by Rupert Murdoch, but not marketized—at least to this date. Second Life, however, is being marketized from within, as it were, through the process of many commercial interests not "buying" the property but buying into the social space.[28]

The upshot is that the analytic distinctiveness of the CIs rests not upon their cultural value or sublime nature (i.e., their nonmarket value) but upon the overarching fact that the environment of both their production and consumption is essentially constituted by complex social networks. The CIs rely, to a greater extent than other socioeconomic activity, on "word of mouth," taste cultures, and "popularity" such that individual choices are dominated by information feedback over social networks rather than innate preferences and price signals. Other people's preferences have commodity status over a social network because novelty, by definition, carries uncertainty and other people's choices therefore carry information.[29]

NOTES

1. See Stanley Metcalfe, *Evolutionary Economics and Creative Destruction* (London: Routledge, 1998); Brian Loasby, *Knowledge, Institutions and Evolution in Economics* (London: Routledge, 1999); Jason Potts, *The New Evolutionary Microeconomics* (Cheltenham, UK: Edward Elgar, 2000).
2. See J. Stanley Metcalfe and Ian Miles, eds., *Innovation Systems in the Service Economy: Measurement and Case Study Analysis* (Boston: Kluwer Academic, 2000); Mark Boden and Ian Miles, eds., *Services and the Knowledge-Based Economy* (London and New York: Continuum, 2000).
3. http://www.mmu.ac.uk/h-ss/mipc/foci/mission.htm.
4. Richard Caves, *Creative Industries: Contracts between Art and Commerce* (Harvard, MA: Harvard University Press, 2000), 1–17.
5. David Throsby and Glen Withers, *Economics of the Performing Arts* (London: Edward Arnold, 1979); David Throsby, "The Production and Consumption of the Arts," *Journal of Economic Literature*, 32 (1994): 1–29.
6. See Ruth Towse, ed., *Cultural Economics* (Cheltenham, UK: Edward Elgar, 1997); Ruth Towse, ed., *A Handbook of Cultural Economics* (Cheltenham, UK: Edward Elgar, 2003).
7. See Caves, *Creative Industries*; Arthur De Vany, *Hollywood Economics: How Extreme Uncertainty Shapes the Film Industry* (London: Routledge, 2004).
8. See Alan Scott, "A New Map of Hollywood: The Production and Distribution of American Motion Pictures," *Regional Studies*, 36(9) (2002): 957–75; Alan Scott, "Entrepreneurship, Innovation and Industrial Development: Geography and the Creative Field Revisited," *Small Business Economics*, 26(1) (2006): 1–24.
9. See Design Council (UK), "The Impact of Design on Stock Market Performance: An Analysis of UK Quoted Companies 1994–2003," February 2004; *Design Taskforce, Success by Design NZ: A Report and Strategic Plan*, Design Taskforce in Partnership with New Zealand Government, in support of the Growth and Innovation Framework (GIF), May 2003.
10. Bengt-Ake Lundvall, *National Systems of Innovation: Towards a Theory of Innovation and Interactive Learning* (New York: Pinter Publishers, 1992); Richard Nelson, "Technology, Institutions and Innovation Systems," Research

Policy, 31 (2002): 265–72; Richard Nelson, ed., *National Innovation Systems: A Comparative Analysis* (Oxford: Oxford University Press, 1993); Chris Freeman, "The National System of Innovation in Historical Perspective," *Cambridge Journal of Economics*, 19 (1995): 5–24; Charles Edquist, *Systems of Innovation: Technologies, Institutions and Organizations* (Washington DC: Pinter, 1997); Johann Peter Murmann, *Knowledge and Competitive Advantage: The Coevolution of Firms, Technology and National Systems* (Cambridge: Cambridge University Press, 2004); Mark Dodgson, David Gann, and Ammon Salter, *Think, Play, Do: Technology, Innovation, and Organization* (Oxford and New York: Oxford University Press, 2005).

11. Loasby, *Knowledge, Institutions and Evolution in Economics*; Chris Freeman, "Continental, National and Sub-national Systems of Innovation," *Research Policy*, 31 (2002): 191–211.

12. Jason Potts and Tom Mandeville, "Toward an Evolutionary Theory of Innovation and Growth in the Service Economy," *Prometheus*, 25(2) (2007): 147–60; Tyler Cowan, *In Praise of Commercial Culture* (Harvard, MA: Harvard University Press, 1998); Tyler Cowan, *Creative Destruction* (Princeton, NJ: Princeton University Press, 2002); Richard Florida, *The Rise of the Creative Class* (New York: Basic Books, 2002); John Howkins, *The Creative Economy: How People Make Money from Ideas* (London: Penguin Books, 2001).

13. Tanja Kotro and Mika Pantzar, "Product Development and Changing Cultural Landscapes—Is our Future in 'Snowboarding'?" *Design Issues*, 18(2) (Spring 2002): 34, 36.

14. Chris Anderson, *The Long Tail* (New York: Random House, 2006).

15. http://www.mindjack.com/feature/piracy051305.html.

16. Henry Jenkins, *Convergence Culture: Where Old and New Media Collide* (New York: New York University Press, 2006); J. C. Herz, "Harnessing the Hive," in *Creative Industries*, ed. John Hartley (Malden, MA: Blackwell, 2006), 327–41.

17. http://www.wethinkthebook.net/home.aspx.

18. Yochai Benkler, *The Wealth of Networks: How Social Production Transforms Markets and Freedom* (New Haven, CT: Yale University Press, 2006).

19. Jason Potts, Stuart Cunningham, John Hartley, and Paul Ormerod, "Social Network Markets: A New Definition of the Creative Industries," *ARC Centre of Excellence for Creative Industries and Innovation* working paper, QUT, July 2007.

20. W. Brian Arthur, "Competing Technologies, Increasing Returns and Lock-in by Historical Events," *Economic Journal*, 99 (1989): 116–31.

21. Arthur De Vany and W. D. Walls, "Bose-Einstein Dynamics and Adaptive Contracting in the Motion Picture Industry," *The Economic Journal*, 106(439) (1996): 1493–1514.

22. Paul Ormerod, *Butterfly Economics* (London: Faber & Faber, 1998) and *Why Most Things Fail: Evolution, Extinction and Economics* (London: Faber & Faber, 2005); "Extracting Deep Knowledge from Limited Information on Evolved Social Networks," *Physica A: Statistical Mechanics and its Applications*, 378(1) (May 2007): 48–52.

23. M. Kretschmer, G. M. Klimis, and C. J. Choi, "Increasing Returns and Social Contagion in Cultural Industries," *British Journal of Management*, 10(1) (1999): 61–72.

24. Jonathan Beck, "The Sale Effect of Word of Mouth: A Model for Creative Goods and Estimation for Novels," *Journal of Cultural Economics*, 31(1) (2007): 5–23.

25. Thomas C. Schelling, "Hockey Helmets, Concealed Weapons, and Daylight Saving: A Study of Binary Choices with Externalities," *Journal of Conflict Resolution,* 17(3) (1973): 381–428.
26. See de Vany, *Hollywood Economics*; Jason Potts, "How Creative Are the Super-Rich?" *Agenda,* 13(4) (2006): 139–50; Beck, "The Sale Effect of Word of Mouth."
27. Pierre Bourdieu, *Distinction: A Social Critique of the Judgment of Taste,* trans. Richard Nice (Cambridge, MA: Harvard University Press, 1984 [1979]).
28. Edward Castronova, *Synthetic Worlds: The Business and Culture of Online Games* (Chicago: University of Chicago Press, 2006).
29. Duncan J. Watts, *Small Worlds* (Princeton, NJ: Princeton University Press: 1999); Peter E. Earl and Jason Potts, "The Market for Preferences," *Cambridge Journal of Economics,* 28 (2004): 619–33.

12 Domesticating New Media
A Discussion on Locating Mobile Media

Larissa Hjorth

INTRODUCTION

As convergence leaves its mark as this century, the ultimate alibi in the convergence rhetoric seems to be the mobile device. Convergence can occur across various levels such as technological, economic, industrial, and cultural. As Henry Jenkins observed, in the growth of mobile phone into converging various forms of multimedia—into the ambiguous and yet ubiquitous mobile media—one could almost forget that mobile media arose from an extension of the landline telephony.[1]

Now the twenty-first century's equivalent to the Swiss army knife,[2] mobile media encompasses multiple forms of media including camera, gaming platform, MP3 player, and Internet portal. As we begin to chart the burgeoning phenomenon of mobile media, we must reassess the methodologies and frameworks being used. How do we grapple with mobile media's interdisciplinary background? Should mobile media be framed in terms of the mobile communication and material cultures traditions, fathered by British theorist Roger Silverstone, that have contextualized the sociocultural processes of media technologies in terms of the domestic technologies approach? Or should mobile media be framed by creative theories and practices of new media?

The rise of the mobile phone into mobile media has attracted scholars from various disciplines such as media studies, gender studies, cultural studies, media sociology, virtual ethnography, and new media, all bringing with them a wealth of traditions, methodologies, and approaches. One of the dominant and highly successful approaches in the field of studying mobile phone cultures is, undoubtedly, the domestic technologies approach.

As an interdisciplinary framework, the domestic technologies approach[3] draws from anthropology,[4] cultural studies,[5] and consumption studies.[6] A significant part of its lineage lies in anthropology and its commitment to analyzing the processes of material cultures in everyday life. Undoubtedly, the seduction of the domestic technologies approach is that it focuses on the symbolic dimensions of technologies in everyday life. In particular, the domestic technologies approach focuses on meanings individuals and cultural contexts give to their technologies, extrapolating on the ways in which users perceive them.

However, as the mobile phone expands into a multimedia device, how can the dimensions of social and reproductive labor—addressed by domestic technologies approaches—be incorporated into the growing realm of mobile media as new media? Domestic technologies approaches seem to fail in grasping the role of creative labor associated with mobile media beyond social and reproductive labor paradigms. In turn, new media approaches to mobile media seem unequipped to address the political dimensions of social and reproductive labor. Since both approaches have been useful in addressing the dynamic, social, creative, and procedural nature of mobile media, it seems fitting to discuss these two enveloping traditions in the context of locating mobile media within "Domesticating new media."

In this chapter I will explore the marriage between the two traditions—on the one hand, the domestic technologies approach, on the other hand, new media remediation approach—in order to conceptualize some of the paradoxes found in mobile media in terms of earlier, ongoing processes. I will outline some of the key attributes and paradoxes that have plagued both traditions' examination of mobile media. Through the example of mobile location-aware gaming, I will draw upon current discourses around mobile media and its coinhabitation in both domestic technologies and new media discourses. As this chapter will argue, through mobile media we can gain insight into some of the recurring paradoxes that run across disciplines and boundaries, continuing to haunt and limit interdisciplinary approaches to twenty-first-century new media practices. In particular, I argue that the emphasis upon visuality and screen-centric views have neglected to address one of the most important aspects of mobile media, the haptic.

MEDIA @ MOBILE

Mobile media is a strange animal to tame. Part domestic technology, part new media, the phenomenon has attached much stargazing and posturing about the future. Through the portal of mobile media, we have witnessed mobility becoming conflated with futurism. The rise of the mobile phone has been marked by its shifting symbolism, usages, and adaptations.[7] When mobile phones first graced the mainstream in the 1980s they were associated with yuppies and conspicuous displays of wealth as demonstrated in the iconic 1980s film *Wall Street*. Then, as mobile phones were adopted and adapted by youth cultures, the phone shrunk into a complex creature adorned by user-created customization from phone straps to sticker faceplates and screen savers. Then, as the phone became more than *just* a phone and started to emanate this century's Swiss army knife, it expanded in size both physically and psychologically to become an integral component in visual, textual, and aural practices in contemporary everyday life.[8]

It is with this size change that we moved into an epoch of mobile multi-modality that became synonymous with contemporary mobility. The rise of mobile media as multimedia par excellence has also been accompanied by corporate smoke and mirrors around the so-called empowered user by way of user-created content (UCC) and prosumer agency. In this climate of optimistic futurism, mobile media promised a further democratization of media. But as Finnish theorist Ilpo Koskinen notes, this accessibility of multimedia often resulted in the aesthetics of banality; images and media rehearse well-known genres and themes.[9] Within the so-called banality are normalized power relations inscribed at the level of everyday practice; thus mobile media serves to remind us of the growing significance of place.

Much work has been conducted around the "banality" of mobile media practices in terms of cameraphone visual and distribution characteristics with many theorists pointing to the content of mobile media rehearsing earlier media (that is, cameraphone images reenacting analogue genres) being banal but the context in which they are shared (or not) providing much signification;[10] however, it seems that the haptic economies, so particular to mobile media, are in need of reevaluation. While Mizuko Ito and Daisuke Okabe's 3 S's—sharing, storing, and saving[11]—noted some of the particulars, we need to examine the politics of "waiting for immediacy" just outside the frame/screen. In other words, what are some of the haptic workouts occurring just outside the frame that undoubtedly affect inside the frame?

So what do I mean by haptic? Just one glance at the current models of mobile media such as iPhone and LG prada and we can see that the screen is no longer about visuality; it is about haptics—haptic screens, to be precise. The engagement of mobile media is not ocular in the case of the gaze or the glance, but rather akin to what Chris Chesher characterizes as the "glaze."[12] Drawing on console games cultures, Chesher identifies three types of glaze spaces—the glazed-over, sticky, and identity-reflective. For Chesher, these three 'dimensions' of the glaze move beyond a visual economy, deploying the filters of the other senses such as aural and haptic.

The haptic has often been undertheorized in mobile communication discourses, often left up to new media practitioners to grapple with in such projects as location-aware gaming. In the growth of mobile-media discourses, much has been discussed in terms of media such as cameraphone practices and the associated sharing and distribution methods. However, much of the rhetoric around mobile media and convergence has been focused upon the frame and visuality—as such concepts as "cross-platforming" entail. These models have discussed media in terms of twentieth-century preoccupations with the visual and the screen, neglecting to reorient frameworks around what makes mobile media so particular; whether being mobile or immobile, the logic is of the haptic. It is about the touch of the device, the intimacy of the object, that makes it so meaningful.

For new media artists such as Rafael Lazano Hemmer and his relational architecture projects, it is this very oscillation of the haptic and the cerebral that partakes in mobile media copresence that makes it such a particular vehicle for twenty-first-century new media practice. In urban spaces, it is not so much the cameraphone images that are transforming the spaces but, rather, the haptic workouts of the everyday user documenting. Much of the discussion of mobile media has encircled the important role of mobile media copresence,[13] and yet the integral notion of the haptic, apart from the hype around SMS thumb cultures, has been largely ignored.

However, the critique of normalized everyday practices and the haptic workouts outside the frame can be found in the various upsurge of experimental new media projects such as location-aware gaming, mobile gaming, or "big games." Location-aware or pervasive games often involve the use of GPS (geographic positioning systems), which allows games to be played simultaneously online and offline. As Finnish theorist (and director of DiGRA) Frans Mäyrä notes, gaming has always involved place and mobility and yet this is precisely what is missing in current games, especially single player genres.[14] Mäyrä points to the possibilities of pervasive (location-aware) gaming as not only testing our imagination and creativity but also questioning our ideas of what constitutes reality and what it means to be copresent and virtual.

The notion of "big games" does not so much relate to the gadget's gluttonous size but rather it has more to do with the role of people and the gravity of place in the navigation of copresence. These projects served to remind us of the importance of locality and its relationship to practices of copresence. The potentiality of "big games" to expose and comment on the politics of copresence—traversing virtual and actual, here and there—in contemporary media cultures has gained much attention. They highlight some of the key paradoxes of everyday life that have been exemplified in mobile-media projects such as location-aware gaming. The paradoxes include virtual and actual, online and offline, cerebral and haptic, delay and immediacy.

As Frank Lantz, a New York–based game designer who has been involved in such pivotal projects as *PacManhattan*, notes, the importance of location-aware mobile gaming—or "big games"—definitely plays an important role in the future of gaming.[15] Citing examples such as *PacManhattan*, UK's blast theory, Geocaching, and Mogi, Lantz emphasizes the importance of these projects in testing the notion of reality as mediation. As Lantz observes, the precursors to big games and the 1970s New Games Movement were undoubtedly the art movements of the 1960s such as happenings (impromptu art events) and the Situationist International (SI) tactics of Guy Debord such as *detourement,* which operated to interrupt/disrupt everyday practices and the increasingly role of media and commodification. In this way, this can be paralleled with the trend in contemporary art from 1990s that French curator and critic Nicolas Bourriaud dubs "relational aesthetics."[16] As Bourriaud observed, "relational aesthetics" dominated the international art scene from the 1990s onwards, building from an emphasis upon locality and deinstitutionalization of installation and the "international" in favor of the vernacular and local.

Locative mobile gaming illustrates the paradoxes of mobile media as part of the cyclic and dynamic processes of technology. For example, in an age of so-called immediate technologies, such projects enlighten us to the conundrum of instantaneity, that is, the inevitable poetics of delay. They highlight the price of mobility and its oscillation between freedom and leash[17] in which work and leisure boundaries are increasingly blurred.[18]

Locative mobile gaming also emphasizes other paradoxes apart from the aforementioned immediacy/delay temporal conundrum. These projects highlight the way in which mobile media can often interfere with, rather than help, face-to-face connections. For example, the tyranny of mobile media's creative labor/democratizing of media dimensions, as epitomized by UCC, sees users becoming more enslaved to the technology rather than it freeing up time to spend with intimates. Locative mobile gaming projects afford us one way in which to reflect and mediate on the paradoxes of contemporary mobile media.

Moreover, locative mobile gaming illustrates that in the face of democratizing of media, new media is still far from the understandings and interests of the everyday person. It also reflects new media artists' fears and yet curiosity about mobile media's ultimate creative conundrum: is it the rise of democratized media and mainstreaming of new media or does the "banality" represent the domination of pedestrianization of new media? Can mobile media teach new media ways to remember the histories "shock of new" as actually the "delay of the banal"?

As a new conflation of many techniques, traditions, and media histories, it is no easy task to outline the nebulous terrain of mobile media. In this chapter I argue that one way in which we can understand mobile media is vis-à-vis its borrowing from, and adapting of, various sociological and new media traditions. In the next section of the chapter I will address two traditions—domestic technologies and remediation, new media approaches—in the cartography of mobile media. I argue that many parallels can be found in the two traditions and that by incorporating the two genealogies we could gain much insight into mobile media.

Just as mobile media needs the rigor of domestic technologies approaches to comprehend the social dimensions of new media, it also needs the innovative approaches of new media theory in order to reconceptualize the conflations between creative and social labor in mobile media's fusion between media communication and new media practices. Through the conflation of domestic technologies and remediation of new media approaches we can begin to conceptualize mobile media as no longer just a "third screen," but, more importantly, a *third space*.

DOMESTICATING NEW MEDIA: TWO EXAMPLES
OF THE MULTITRADITIONS OF MOBILE MEDIA

The rise of mobile media could be read as nascent. However, such a belief, propagated in global media's lauding of the new mobile revolution in consumer agency (in the form of the prosumer and Web 2.0), neglects to address the dynamic dimensions of technology as a sociotechnological process. In the case of domestic technologies approaches, in which domestication is always an ongoing and never-completed process, the dynamics of mobile media extends already existing cyclical models. So too, in the tradition of new media, in which old and new have had a dialectical and dynamic relationship that disrupts any linear or casual notion of time.

As mobile communication and media industries converge, the all-pervasive futurist rhetoric becomes stifling. And yet, if the twin histories of new media and mobile communication have taught us anything, the "new" is always remediated and mediated. Each "new" technology deploys techniques of the older technology, which in turn revises the earlier media. This cuts to the core of all communication and cultural practices implicated in intimacy. For Jay Bolter and Richard Grusin, new media are remediated with older media into a dynamic ongoing process that disrupts any causal or linear notion of old and new technologies.[19] As Margaret Morse concisely notes in the case of the Internet, all forms of intimacy are mediated—by language, gestures, and memories.[20] Emerging forms of visual, textual, and haptic mobile genres such as SMS and cameraphone practices—reenacting earlier rituals such as nineteenth-century letter writing, postcards,[21] and gift-giving customs[22]—have only served to highlight the remediated nature of the rise of mobile media.

There is much to be learnt from understanding the parallels between new media theory on remediation and mobile communication's usage of the domestic technologies approach. Like the domestic technologies approach,[23] the study of new media through the lens of remediation echoes a similar philosophical stance. As influential theorist in the field of media archaeology, Erkki Huhtamo, has argued, the cyclical phenomena of media tend to transcend historical contexts, often placating a process of paradoxical reenactment and reenchantment with what is deemed as "new."[24] For Huhtamo, media archaeology approaches are "a way of studying recurring cyclical phenomena that (re)appear and disappear over and over again in media history, somehow seeming to transcend specific historical contexts."[25] As Jussi Parikka and Jaakko Suominen note, the procedural nature of media archaeology approaches means "new media is always situated within continuous histories of media production, distribution and usage—as part of a longer duration of experience."[26]

Citing an example of the launch of Nintendo DS that heralded a new and "unique" experience for twenty-first-century entertainment, Parikka and Suominen note that much of contemporary postindustrial digital media culture is inundated by futurism that seeks to break with the past.[27] Parikka and Suominen note that this "creates the impression that, in the new media discourse, the past functions solely as something worse or less sophisticated, something that has to be left behind and practically forgotten."[28]

When Marshall McLuhan identified that the content of new media is that of the previous technology, he highlighted the nonlinear and dynamic role of new technology imbued by the specters of old technology.[29] In short, that the "new" is far from superseding or breaking with the old as modernist mythologies would have it. The fact that the notion of "new" in new media has been continuously challenged and demonstrated as a fallacy echoes the way in that technology has been approached by many mobile communication scholars (from predominantly sociological and urban anthropological traditions) through the domestic technologies approach.[30]

In the picture painted by the domestic technologies approach, domestic technologies such as the radio, TV, and mobile phone are seen as part of the cyclic and ongoing process of consumption in everyday contemporary life. As Daniel Miller notes in his coauthored study with Heather Horst on Jamaican cell-phone use, "what one has to study are not things or people but processes."[31] Much of the literature analyzing mobile communication has utilized the domestic technologies approach[32] to identify adoption and adaptation of technologies as always ongoing and never completed.[33] Like the cultures in which they inhabit, domestic technologies are always in flux. Domestication is ongoing and dynamic, and through customization practices we can domesticate domestic technologies as much as they domesticate us in a productive tension. In the case of the mobile phone, while the domestic technology device may have *physically* left the home, it *psychologically* resonates what it means to be at home and local no matter where it is located.

As David Morley has noted, the mobile phone has often been cited as a key example of domestic technologies par excellence.[34] Key scholars in this area include Leslie Haddon, Roger Silverstone, Haddon and Silverstone, Rich Ling, and Miller. As a key scholar in the field, Haddon provided decisive apparatus to comprehend the dynamic and enduring processes of the domestic technologies.[35] Often users' relationships to their domestic technologies can wax and wane, drawing feelings of ambivalence, and yet inevitably due to the prescribed need to have the technologies to be part of contemporary urbanity.

As an approach, the domestic technologies method sees the process of engagement with technologies undergoing various stages or nodes of a cycle that include "imagination, appropriation, objectification, incorporation, and conversion."[36] Consumption is seen as an ongoing process that is perpetually negotiated, way after the actual point of sales. As Rich Ling notes, "our consumption becomes a part of our own social identity.

Further, others' consumption is a type of lens through which we see them and through which we interpret their social position."[37]

Mobile media represents a meeting of the crossroads between the genealogy of domestic technologies and media archaeologies of new media. In both these traditions, we see mobile media remediating and reenacting previous media cultures and modes of domestic regimes. Reminding us of our forgetting whilst harnessing the inevitable amnesia that accompanies any notion of "new," mobile media represents the conundrum of new technologies. In new media discourses we can find many examples of the content or specters of the older media. Like the domestic technologies approach, the study of new media through the lens of remediation echoes a similar philosophical stance.

According to Timo Kopomaa, the mobile phone is an extension of nineteenth-century media.[38] For Kopomaa, mobile media creates a new "third" space in between public and private space. On the one hand, the project of examining mobile media entails observing the remediated nature of new technologies and thus conceptualizing them in terms of media archaeologies.[39] On the other hand, mobile media's reenactment of earlier technologies is indicative of its domestic technologies tradition that extends and rehearses the processes of precursors such as radio and TV. It is the fact that Kopomaa draws our attention to mobile media as a third space, rather than third screen, which is significant.

Both traditions—the domestic technologies and new media remediation approaches—emphasize the cyclic and dynamic process of media technologies that cannot be simplistically divided between old and new or inside and outside the screen. Rather, the cartography of mobile media is one imbued by paradoxes. In the case of cameraphone practices—whether still or moving—mobile media demonstrates two distinctive paradoxes, that of the *reel* in the real, and the inherent poetics of *delay* in the practice of immediacy in the navigating of offline and online copresence. As Lev Manovich identified,[40] contemporary new media and digital practice are all consumed by fetishizing the real through the lens of the reel—that is, texture and skin of the analogue.

For Manovich, the way in which to understand the remediated emerging digital cultures and the haunting by the ghost of the analogue is through a series of paradoxes. These sets of paradoxes are located around the relationship between the real and the reel. As Manovich identifies, while the analogue may disappear, it will continue to haunt the digital in the form of the analogue's particular realism, the "reel." This is evident in the way in which cameraphone practices echo previous analogue norms[41] and that, in turn, make mobile media, according to Koskinen, characterized by "banality."

However, one of the most compelling examples of the real/reel phenomenon, where the tactile process of the analogue is fully felt both metaphorically and actually, is the rise of screen cultures in mobile media. In particular, the rise of such mobile-media devices as iPhone, LG prada, and

Samsung Arami phone—to name a few—all incorporate one key feature, haptic screens. Here the reel/real paradox is played out in the haptic versus visual, in which the haptic is undoubtedly the more meaningful factor that "domesticates" the device into the user's everyday life. Much of the specters of the analogue reel are more about the tactical experience of image processing; and while these processes have been deleted in the rise of the digital, it is the legacy of the haptic—that has moved from the filmic developing process to the actual politics of the touch screen—that continues unabated. However, in the language set of twentieth-century media cultures, much discussion was given to visuality rather than the increasing role of the haptic.

While location-aware projects are invaluable in geocaching (such as GPS) and demonstrating the importance of place and specificity in a period of global technologies, they also served to highlight one of the greatest residual paradoxes of mobile media as a metaphor for sociotechnologies. That is, the paradoxical politics of copresence. One example can be found in the aims of twentieth-century technology to overcome difference and distance from geographic and physical to cultural and psychological. This attempt to overcome distance and difference sees the opposite result, the overcoming of closeness. Practices of copresence intimacies become fetishized, through what Misa Matsuda has characterized as "full-time intimacy."[42] This recites what Michael Arnold identified as the Janus-faced nature of mobile media that operates to push and pull us, setting us free to roam and yet attaches us to a perpetual leash.[43]

As Arnold notes, the Janis-faced phenomenon is symptomatic of what Martin Heidegger characterized as "un-distance." The role of technology in the twentieth century has always been to overcome some form of distance—whether geographic, physical, social, cultural, temporal, or spatial. But herein lies the paradox. The more we try to overcome distance, the more we overcome closeness. This is the kernel of un-distance and its temporal and spatial tenor. Un-distance can be seen today in the practice of mobile media, particularly pervasive location-aware projects that rely on the so-called immediacy or instantaneity of the networked.

However, one could argue that un-distance has been perpetuated by the ocular-centrism of twentieth century "tele" media, a phenomenon that has been disrupted by mobile media's emphasis on the haptic. For Ingrid Richardson, mobile media needs to harness the importance of the haptic. Conducting a small ethno-phenomenological study on the use of phonegame hybrids, Richardson disavows the ocular-centrism prevalent in "new media screen technologies" to focus on "the spatial, perceptual and ontic effects of mobile devices as nascent new media forms."[44] As she persuasively observes,

> In order to grasp the epistemic, ontic and phenomenological status
> of screen media it is important to trace their ocularcentric legacy; by

understanding this history we can then interpret how mobile screens in particular work to bewilder classical notions of visual perception, agency and knowing.[45]

Indeed, one of the compelling factors to arise from mobile media, and this links back to its fusion of remediation and domestic genealogies, is the persistence of the ontology of the reel. However, unlike the twentieth-century "reel"—in the form of the aural modes of address embroiled in "screenness"—the mobile reel, and thus possible creative worlds and realities, is undoubtedly governed by the haptic.

The game of mobile media—whether it be partaking in cameraphone imagery and the haptic exercises outside the screen, to mobile gaming, in which interactivity and engagement are navigated by haptic mobility and immobility rather than visualities of screen cultures—is undoubtedly changing how we are thinking about domestic technologies and new media. Through the lens of paradoxes that encompass virtual and actual, online and offline, haptic and visual and delay and immediacy, some lessons about twentieth-century media practice can be learnt. For anyone that has participated in a mobile pervasive game, they will quickly identify the lack of coherence between online and offline copresence. The more we try to partake in the *politics of immediacy,* the more we succumb to the *poetics of delay.* This paradox extends beyond just mobile gaming and can be found in many of the multimedia possibilities of mobile media—from cameraphone imagery and MMS to moblogging and SMS.

In the case of the growing interest in urban screen cultures as an analogy for the twenty-first century, one could argue that it is indeed the very eruption of the twentieth century's obsession with visualities for the twenty-first century's politics of the haptic that dominates the canvas of mobile media. From the haptic screen interfaces to the various multimedia tactics such as mobile gaming that disavow the screen for the haptic and audio, mobile media revises the perceived hierarchies of the senses which, in turn, could breath new life into new media and domestic technologies approaches.

TIME AFTER TIME: THE NEVER-ENDING CONCLUDING BEGINNING

In modernism, the role of originality was celebrated. For the modernist avant-gardists, such vehicles as technology served as a decisive break from the past in what art critic Robert Hughes characterized as "the shock of the new."[46] In contemporary postindustrial digital cultures, the "new" promised by mobile media is in fact "banal" and located in nostalgic politics such as the real/reel paradigm. In this chapter, I have assembled two traditions—domestic technologies and remediation—in order to show the similar cyclic debates operating across disciplines. I argue that perhaps mobile media needs to be conceptualized as the "shock of the banal," that is, its paradoxes—online and offline, virtual and actual, delay and immediacy, haptic and visual—are far from new and can be traced through various disciplinary traditions.

In my ethnographic studies into cameraphone practices in Seoul, Tokyo, Hong Kong, and Melbourne, one of the increasing features of the tyranny of the "full–time intimacy"[47] of mobile media customization is the use of immediacy to camouflage delay. Many respondents spoke of creating their own forms of delay so that they could savor the SMS or MMS. With the immediacy of such technologies, the tactics of pretending to not see or receive a message immediately allowed respondents time—what I call "the poetics of delay." Moreover, the persistence of the reel in much of the cameraphone images, genres, and mobile movies was significant. In particular, for many respondents, mobile-media making was less about visual economies and more about aural and haptic modes of address akin to earlier "reel" domestic technologies such as the TV and radio. Thus, the conflation between domestic technologies and new media approaches could further address one of the greatest paradoxes of shifts from twentieth to twenty-first century media; that is, rather than it being a history predicated on visualities as the "screenness" would entail, it is a history of the rise of the audio and the haptic that are becoming the key indicators and characteristics of mobile media.[48]

I have chosen to focus on two traditions—one draws from media and communication and material cultures (anthropology) in the form of the domestic technologies approach; the other calls upon new media approaches of remediation and media archaeologies approaches. In these two traditions we can see various similarities—the focus on the dynamic, sociocultural processes of mobile media. While the former allows for more insight into social and reproductive labor debates, the latter affords us acuity into shifting modes of accessing creative labor and everyday life. In the case of mobile media projects such as location-aware gaming, I argue we need to draw upon the two models, incorporating them into a new framework for evaluating dimensions of mobile media, and twenty-first century screen cultures, in terms of key attributes such as the haptic. The important factor here is against the seductive and simplistic futurism prevailing in much discussion around mobile media we need to recognize that mobile media, like new media, is inevitably involved in the politics of the banal and nostalgia.

Much of the futurist posturing accompanying mobile-media discussions in global media have celebrated the potential democratization of new media. With the rise of the prosumer from the term coined by Alvin Toffler in 1980[49] to its adaptation by Don Tapscott in 1995[50] to the context of the Internet's digital economy, much of the media of late has celebrated UCC and the prosumer as part of the Web 2.0 enterprises. But behind this rhetoric is the pivotal role mobile media has played in creating and reenacting debates about technology, labor, and creativity that have long accompanied new media and domestic technology discourses. Rather than just domestic and artistic labor having little or no remuneration in the general community, now the UCC associated with mobile media could see the everyday person subject to the injustices of industry convergence whereby corporations buy and exploit social and creative labor in the form of Web 2.0 media such as SNS.

The conundrum of new mobile technologies is that they are *supposed* to free us up and yet, as a good existential crisis would have it, the freedom is a leash. Work becomes mobile; labor is on a perpetual drip. We are supposed to be available at all times, perpetually connected. Rather than free us, the "immediacy" logic of mobile technologies makes us feel like we must be quicker and must achieve more. Rather than saving time, applications such as cameraphone image making—and the attendant customizing and modes of sharing/distribution—mean users spend a lot of time sharing and editing the so-called immediate. The present gets put on hold. However, one of the features that becomes apparent in mobile media is the need to move beyond the screen-centric and ocular-centricism of twentieth-century media and reconnect with the very reason the mobile phone has grown into mobile media, its importance at the level of everyday haptics.

By engaging in the significance of the haptic in mobile media we can grasp some of the paradoxes at play. It is important to recognize that this conundrum of delay and immediacy is not new with the rise of mobile media. Rather, these paradoxes have been central in the emphasis upon screen cultures in the face of the importance of the haptic in the making sense of mobile media. As I have attempted to discuss, mobile media represents some interesting paradoxes about contemporary media and consumer cultures. In this chapter I have tried to show the ambivalences surrounding mobile media from both a new media and domestic technologies approach in order to reconceptualize the philosophical and phenomenological dimensions of mobile media. To socialize the creative media dimensions and to innovate the social, domestic dimensions.

In order to grapple with the burgeoning field of mobile media we need to comprehend the twin histories—such as the domestic technologies and remediation approach—to fully grasp the histories, contemporary and future paradoxical permutations of mobile media and not just fetishize the "new" by futurist posturings. Mobile media is undoubtedly a project involving the domesticating of new media in which old boundaries between art and life, production and consumption perpetually change and shift, repeat and pause. But it is time that we moved away from the twentieth-century preoccupations with visual cultures and screen-ness that deems to view mobile media as a (advertising) *third screen* and instead acknowledge its genealogy as a *third space* that is governed by the politics and aesthetics of haptics.

NOTES

1. Henry Jenkins, "Welcome to Convergence Culture," *Receiver*, 12 (2005) http://www.receiver.vodafone.com/12/articles/pdf/12_01.pdf.
2. John Boyd, "The Only Gadget You'll Ever Need," *New Scientist*, 5 (2005): 28.
3. Roger Silverstone and Eric Hirsch, eds., *Consuming Technologies: Media and Information in Domestic Spaces* (London: Routledge, 1992); Roger Silverstone and Leslie Haddon, "Design and Domestication of Information and Communication Technologies: Technical Change and Everyday Life," in *Communication by Design: The Politics of Information and Communication Technologies*, ed. Roger Silverstone and Richard Mansell (Oxford, UK: Oxford University Press, 1996): 44–74; Daniel Miller, *Material Culture and Mass Consumption* (London: Blackwell, 1987).
4. Mary Douglas and Baron Isherwood, *The World of Goods: Towards an Anthropology of Consumption of Goods* (London: Routledge & Kegan Paul, 1979).
5. Dick Hebdige, *Hiding in the Light: On Images and Things* (London: Routledge, 1988).
6. Miller, *Material Culture and Mass Consumption*.
7. John Agar, *Constant Touch: A Global History of the Mobile Phone* (Cambridge: Icon Books, 2003).
8. Boyd, "The Only Gadget You'll Ever Need."
9. Ilpo Koskinen, "Managing Banality in Mobile Multimedia," in Raul Pertierra, ed., *The Social Construction and Usage of Communication Technologies: European and Asian Experiences* (Manila: University of the Philippines Press, 2007), 48–60.
10. Barbara Scifo, "The Domestication of the Camera Phone and MMS Communications: The Experience of Young Italians," in Kristóf Nyíri, ed., *A Sense of Place: The Global and the Local in Mobile Communication* (Vienna: Passagen Verlag, 2005), 363–73; Mizuko Ito and Daisuke Okabe, "Camera Phones Changing the Definition of Picture-Worthy," *Japan Media Review* (2003), http://www.ojr.org/japan/wireless/1062208524.php; Mizuko Ito and Daisuke Okabe, "Intimate Visual Co-Presence," presented at *UbiComp 2005*, 11–14 September, Takanawa Prince Hotel, Tokyo, Japan, http://www.itofisher.com/mito/.
11. Ito and Okabe, "Intimate Visual Co-Presence."
12. Chris Chesher, "Neither Gaze nor Glance, but Glaze: Relating to Console Game Screens," *SCAN: Journal of Media Arts Culture*, 1(1) (2004), http://scan.net.au/journal/.
13. Ito and Okabe, "Intimate Visual Co-Presence."
14. Frans Mäyrä, "The City Shaman Dances with Virtual Wolves—Researching Pervasive Mobile Gaming," *Receiver*, 12 (2003), www.receiver.vodafone.com.
15. http://www.pacmanhattan.com/.
16. Nicolas Bourriaud, *Relational Aesthetics*, trans. Simon Pleasance and Fronza Woods (Dijon, France: Les Presses du Réel, 2002).
17. Michael Arnold, "On the Phenomenology of Technology; the "Janus-faces" of Mobile Phones," *Information and Organization*, 13 (2003): 231–56.

18. Judy Wajcman et. al, "Intimate Connections: The Impact of the Mobile Phone on Work Life Boundaries," see this volume. Also see Melissa Gregg, "Work Where You Want: The Labour Politics of the Mobile Office," presented at *Mobile Media Conference* (University of Sydney, July 2007).
19. Jay Bolter and Richard Grusin, *Remediation: Understanding New Media* (Cambridge, MA: MIT Press, 1999).
20. Margaret Morse, *Virtualities: Television, Media Art, and Cyberculture* (Bloomington: Indiana University Press, 1998).
21. Larissa Hjorth, "Locating Mobility: Practices of Co-presence and the Persistence of the Postal Metaphor in SMS/MMS Mobile Phone Customization in Melbourne," *Fibreculture Journal*, 6 (2005), http://journal.fibreculture.org/issue6/issue6_hjorth.html.
22. Alex Taylor and Richard Harper, "Age-Old Practices in the "New World": A Study of Gift-Giving between Teenage Mobile Phone Users," in *Changing Our World, Changing Ourselves* (proceedings of the *SIGCHI* Conference on Human Factors in Computing Systems, Minneapolis, 2002): 439–46; Alex Taylor and Richard Harper, "The Gift of Gab? A Design Oriented Sociology of Young People's Use of Mobiles," *Journal of Computer Supported Cooperative Work*, 12 (2003): 267–96.
23. Douglas and Isherwood, *The World of Goods*.
24. Erkki Huhtamo, "From Kaleidoscomaniac to Cybernerd: Notes Toward an Archaeology of the Media," *Leonardo*, 30(3) (1997).
25. Erkki Huhtamo, "From Kaleidoscomaniac to Cybernerd," 222; cited in Jussi Parikka and Jaakko Suominen, "Victorian Snakes? Towards a Cultural History of Mobile Games and the Experience of Movement," *Games Studies: The International Journal of Computer Game Research*, 6(1) (2006), December, http://gamestudies.org/0601.
26. Parikka and Suominen, "Victorian Snakes? Towards a Cultural History of Mobile Games and the Experience of Movement."
27. Parikka and Suominen, "Victorian Snakes?"
28. Parikka and Suominen, "Victorian Snakes?"
29. Marshall McLuhan, *Understanding Media* (New York: Mentor, 1964).
30. Haddon and Silverstone, "Design and Domestication of Information and Communication Technologies"; Miller, *Material Culture and Mass Consumption*.
31. Daniel Miller and Heather Horst, *Cell Phone* (Oxford and New York: Berg, 2006), 7.
32. Rich Ling, *The Mobile Connection* (San Francisco: Morgan Kaufmann Publishers, 2004).
33. Leslie Haddon, *Empirical Research on the Domestic Phone: A Literature Review* (Brighton, UK: University of Sussex Press, 1997).
34. David Morley, "What's 'Home' Got to Do with It?" *European Journal of Cultural Studies*, 6(4) (2003): 435–58.
35. Leslie Haddon, "Domestication and Mobile Telephony," in *Machines That Become Us: The Social Context of Personal Communication Technology*, ed. James E. Katz (New Brunswick, NJ: Transaction Publishers, 2003), 43–56.
36. Ling, *The Mobile Connection*, 28.
37. Ling, *The Mobile Connection*, 27.
38. Timo Kopomaa, "The City in Your Pocket," in *Birth of the Mobile Information Society* (Helsinki: Gaudemus, 2000).
39. Huhtamo, "From Kaleidoscomaniac to Cybernerd."
40. Lev Manovich, "The Paradoxes of Digital Photography," in *The Photography Reader*, ed. Liz Wells (London, Routledge, 2003), 240–49.

41. Lisa Gye, "Picture This," paper presented at *Vital Signs* conference (September 2005, ACMI, Melbourne).
42. Cited in Mizuko Ito, Daisuke Okabe, and Misa Matsuda, eds., *Personal, Portable, Pedestrian: Mobile Phones in Japanese Life* (Cambridge, MA: MIT Press, 2005).
43. Arnold, "On the Phenomenology of Technology."
44. Ingrid Richardson, "Pocket Technoscapes: The Bodily Incorporation of Mobile Media," in *Continuum: Journal of Media and Cultural Studies*, 21(2) (2007): 205.
45. Richardson, "Pocket Technoscapes," 208.
46. Robert Hughes, *The Shock of the New* (London: Thames and Hudson, 1981).
47. Matsuda, cited in Ito, *Personal, Portable, Pedestrian.*
48. Richardson, "Pocket Technoscapes."
49. Alvin Toffler, *The Third Wave* (William Morrow: New York, 1980).
50. Dan Tapscott, *The Digital Economy: Promise and Peril in the Age of Networked Intelligence* (New York: McGraw-Hill, 1995).

Part IV
Renewing Media Forms

13 Back to the Future
The Past and Present of Mobile TV

Gabriele Balbi and Benedetta Prario

The aim of this chapter is to illustrate the historical developments of two technologies in Italy, namely, Araldo Telefonico in the first decade of the twentieth century and mobile TV in the twenty-first century. These two media may be somewhat different, on an economic and social-technological level, but they also have a common key factor: the impossibility of distinguishing between point-to-point communication and broadcasting.

Araldo Telefonico and mobile TV are two examples of the transition from telecommunications to mass media. Even if we will suggest that the telephone had both a point-to-point and broadcasting nature, fixed and mobile telephony were both originally two typical examples of tele-communications: in fact they were tools created and used by people to communicate with each other. Later, the fixed and mobile phone evolved into different forms and especially into two mass media (Araldo Telefonico and mobile TV), often maintaining the same device (the telephone), habits, and sometimes the same operators, too. In fact, telecommunications operators are now investing in mobile TV.

Furthermore, even if Araldo Telefonico belongs to a different historical period from mobile TV, it achieved similar forms of media convergence (telecommunications and editorial content), of interactivity, and of confusion between the private and public sphere (it was listened to in homes, hotels, bars, and so on): all features in common with the mobile TV, but conceived between the nineteenth and twentieth centuries.

BETWEEN ME, YOU AND *US*. CIRCULAR TELEPHONY IN ITALY

The telephone was invented either as a point-to-point or as a broadcasting medium. Philippe Reis, who coined the term *telephone* in the 1850s, didn't imagine a tool of personal communication; instead, his experiments were conducted with a fairly large audience listening to an artist singing in a sealed room.[1] Even Alexander Graham Bell saw the telephone not as a medium for "establishing direct communication between any two places in the city only, but also as an electrical toy, for broadcasting music."[2] So, since its invention, the telephone has not been regarded solely as a person-to-person communication medium; its inventors also tried to spread "broadcasting uses" such as news diffusion, entertainment shows, and editorial content. From the beginning of its history, in the logic and in the possible uses of the telephone, both concepts were assumed by its inventors and users.

This assumption could be validated in many countries and cultures, but in this chapter we will focus specifically on the telephone in the Italian context. In Italy, the telephone arrived at the beginning of the 1880s and was immediately used to transmit entertainment shows. One of the first official experiments of long-distance telephony in Italy linked the telegraph office of Tivoli and the Quirinale in Rome; "it began with the royal anthem, played on a piano in the Tivoli office. Then some pieces by Prati, a tenor *romanza*, flute and violin sonatas and a poem followed."[3]

This feature also emerged in an article which appeared in the popular Italian newspaper *L'Illustrazione Italiana* on January 29, 1882, which recalled a peculiar experiment carried out by the Gerosa brothers, pioneers of telephony in Milan: from time to time, they played an accordion in the switching room and connected it to all their subscribers, who could thus hear a sonata.[4]

While the Gerosa brothers' sporadic surprises could be considered the first steps of Italian broadcasting, from the beginning of the 1880s in other European countries and in the United States these uses of the telephone were institutionalized with so-called circular telephony. The first experiment of this "novelty" took place in Paris in 1881 during the International Electric Exhibition. But the most popular example of the "Théâtrophone" was launched and developed in the metropolitan area of Budapest by Theodor Puskas in 1893: the so-called Telephone Hirmondò with over 9,000 subscribers. The system was fairly easy to implement. Through dedicated electrical circuits, similar to those of the telephone, subscribers could receive a single "channel" with daily news and shows from a transmitter. When a new "telephone hearing" was organized, the telephone's ring was different from the standard one: a few seconds later the show could begin, following an agenda already known to those who paid for the service: what today is called a (radio or TV) schedule. So the audience could listen to their favorite programs peacefully sitting at home, and it was the first time in history in which premade entertainment shows were directly aimed at the home. The Hirmondò model was copied in many countries:[5] from the English "Electrophone" to the American "Telephone Herald," from the long-lived Soviet example to the Italian "Araldo Telefonico" (Herald and Araldo are quite simply literal translations of the Hungarian "Hirmondò").

In Italy, a previous attempt to formalize entertainment transmissions through telephone lines was made in 1887, as recalled by *Il Telegrafista*, a technical journal of that time written for and read by electrical engineers,[6] but a regular telephone broadcasting service appeared only around 1910.

With a ministerial decree of May 22, 1909, an Italian engineer, Luigi Ranieri, obtained a grant to create a "speaking journal" in Rome: "Araldo Telefonico" was born, inspired by the Hungarian model.

Until the first years of the war, in fact, the Italian Herald copied the Hirmondò schedule: news, various shows and emissions above all from theatres in Rome, popular programs such as the weather forecast, foreign-language lessons, and, especially, the time signal that represented a genre in early Italian radio broadcasting too.[7]

Besides the abovementioned function modalities, an advertising brochure from the early twentieth century underlined the critical distinction between the "speaking journal" service and the telephonic one: in fact, Araldo's subscribers were not obliged to pay a subscription to the telephone too: the two networks were completely independent. There could be many reasons for this. Firstly, the limited diffusion of the telephone in Italy: if telephone subscribers were the only ones allowed to receive programs, there would have been a large number of people who, albeit interested in this service, would have been unable to pay the subscription (Araldo was much cheaper than the traditional telephone). The "Araldo Telefonico" managers probably did not want to relate their technologies to another one that seemed to have diffusion troubles and was owned by the richest people only. Secondly, there was a technical reason, too: Araldo's network had to be distinguished from the telephone's. Telephone networks are built to allow one person to communicate with another one; on the other hand, in Araldo's networks there is only one subject that has to communicate, or better, to spread information, to all the others (broadcasting).

Before the First World War, service costs were basically of three types. The most expensive was the subscription: 60 lire per year (5 per month) for a system installed within a kilometer of the Colonna Square central bureau. Then, there were two *una tantum* (one-off) costs: the first (16.50 lire) to set up the system and the second (20 lire) to guarantee the technologies.[8] Figures show that in 1914 there were 1,315 subscribers to Araldo in Rome. After the First World War, Milano and Bologna offices were opened, and in Rome the "Araldo Telefonico," now directed by Luigi Ranieri's son, slowly transformed into a radio broadcasting system between 1923 and 1924.

An emblematic episode for our paper is the law proceedings involving Luigi Ranieri. Ranieri was a debtor in arrears for a long time and in his defense—and in the judicial uncertainties—a peculiarity of the early twentieth-century mentality can be seen: the impossible distinction between point-to-point and multicast communication. In an article that appeared in the *Rivista delle Comunicazioni* in June 1917 (whose title was significant: "Araldo does not represent telephonic conversation"), a Rome law-court judgment passed down on April 11, 1917, says: this pronouncement closed a very long quarrel between the father of circular telephony and the government.

For personal and financial reasons, the engineer Ranieri could not pay the fees for his business so the ministry revoked the grant with an ordinance on February 20, 1914, and then, on September 22, it ordered him to interrupt all transmissions. With an action repeated by the Italian private TV networks sixty years later,[9] and "despite the revocation of the grant and the injunctions, 'Araldo telefonico' carried on transmitting abusively. Six times the administrative employees interrupted the Araldo service and every time Mr. Ranieri restored it."[10] On June 5, 1916, the chief magistrate in Rome found against Mr. Ranieri, but he appealed claiming "the absence of the misdemeanour."[11] On May 18, 1916, the county council acknowledged "the peculiarity of the service and the substantially different aim from that of the normal telephone grantees."[12]

It is useful to remember the motivations of the sentence, mainly, where it becomes clear that telephone laws and rules cannot be applied to Araldo. For the first time, the distinction between point-to-point and mass communication is made in law and, *tout court*, in contemporary mentality:

> *Communicating*, actually, is different from *transmitting*, because it implies the possibility of responding: this is what Araldo, due to its technology, could not do. It transmitted every day, at fixed times, political news and music collected by Ranieri; it resembled the newspaper rather than the telephone. Mr. Ranieri broadcast the news not through the press, but using technology founded on the principles of telephony: but that system was not the telephone in the common sense of the word. So neither communication nor telephone. . . . the appellant must be absolved from this infringement because his action cannot be considered a crime. In fact, it lacks one of the essential characteristics for being punishable: the objective nature of telephone communication, that in the Araldo does not appear.[13]

"Araldo Telefonico" did not follow the principles of use of traditional telephone communication, so it could not follow its laws. The perception of this distinction appeared in the period 1910–1920 and this is very important: until this moment, in fact, the dissimilarity between interpersonal and broadcasting media could not even be imagined.

This is what an Italian historian, Peppino Ortoleva, underlines about the tele-vision idea. Until the 1920s, this notion had fluctuated between two different concepts too: that of video-telephony (peer-to-peer) and TV in the common sense (broadcasting).

Only after 1920 could video telephony be thought of as a technology in its own right, distinct from television. In fact, a conceptual distinction had to be imposed, obvious for us but not at that time: the distinction between point-to-point and mass communication. In the years 1880–1920 audiovisual communication was a confused area, but basically unified: the radio was point-to-point and mass; the telephone too . . ., telephotography too. It could be presumed that television would be too.[14]

As shown by Ranieri's legal case, the distinction perceived between personal and broadcasting communication could probably be located around the period 1915–1920. But Ortoleva's theory about a confusion of these two concepts (for the tele-vision) is corroborated also in the case of fixed telephony.

It must still be remembered that the "pleasure telephone" era, as the historian Asa Briggs[15] defined this early period of the technology, ended with the advent of the radio. In fact, from the twenties a hiatus appeared in the media system: on one side, some media (for example, the telephone) were definitely associated to point-to-point communication; on the other, new media like the radio and then TV adopted a mass diffusion model: so-called broadcasting. That distinction was appropriate until the advent and commercialization of a new medium: mobile TV.

DEFINITIONS OF MOBILE TV: ITS ADOPTION AND SUPPLY

Over the last ten years digital technology has encouraged rapid growth in the personal consumption of media. The advent of personal video recorders (PVRs), video-on-demand, and the multiplication of programs have enabled viewers to personalize the content they want to watch. And, with interactivity, viewers can directly express their preferences to broadcasters.[16] As part of this trend, and alongside the growth of mobile telephony, "the place of viewing is no longer limited to the television receiver at home, or in a vehicle, but is widened to allow personalized viewing of television by individuals wherever they are located."[17]

So mobile television means the possibility of viewing traditional television and interactive programs directly on one's own mobile. Interactivity is an essential feature of "broadcasting on the move," because this new TV model is closely related to the telephone, the preeminently interactive communication medium. Mobile television can be defined as the possibility of watching television programs on a handheld device and "on the move"—in public transport, waiting for an appointment, or while at work.

The idea of watching television while on the move and not at home is not new. As was well illustrated by Trefzger,[18] in 1982 Sony introduced its first portable television, the Watchman; but unlike its music peer, the Walkman, it did not have much success. Nowadays consumers use their mobile phones for multimedia not just for communication but also for entertainment and for news and information services. With your mobile you can do a whole range of things, and today it seems to be the turn of television. Television and the mobile are facilities that most people cannot imagine living without.

Over the years, a lively scientific debate has attempted to bring media analysis back into the context of people's everyday lives. Fundamental media sociology concepts, like *domestication* discussed by Roger Silverstone,[19] have been defined in order to give greater importance to processes social subjects use in building their own media experience. The more recent tradition of sociocultural studies of information and communication technologies (ICT) prompts us to adopt new research approaches towards the media and, in particular, towards the new media. An interesting branch of media research has, in the last few years, reevaluated the role of the user (*user studies*). The starting point is the sociology of technology, which generically refers back to the "social shaping of technology" approach. The combination of this approach, the cultural audience and media studies, the social history of technology, the sociology of everyday life, and the sociology of consumption partially shift attention from media production to media use and consumption. In this perspective, the pertinent level for investigating the social uses of the media is not only the level of concrete practices but also that of the meanings attributed to the media by people and by the associated individual and social representations.

As shown by experience with innovations, technology alone is not the only determining factor for the success of a product. Above all, for products and services developed for consumers, user acceptance is a critical success factor. The first studies on how families adopt new products and/or innovative services were conducted by Everett Rogers[20] and Frank Bass,[21] but in recent years many case studies have appeared in the economic literature on the adoption of specific mass media, like color television and the video recorder, with the aim of defining a model. In the field of mobile applications the relevance of user acceptance is high, too.[22]

Similar to Roger's Diffusion Theory, it is possible to segment European mobile-phone users into three clusters: Fanatics; Followers; and Fugitives. *Fanatics* always use up-to-date technology and own the newest handsets. They are keen to try out new products and services and are also interested in the technology behind them. Sixteen percent of mobile users belong to this cluster. *Followers* represent 42 percent of the population of users. They appreciate the practical usefulness of technology and adopt it when they believe it is mature. *Fugitives*, who prefer easy usability, represent the final 42 percent of users. They do not adopt new technologies until it is absolutely necessary.

With reference to mobile TV, one can see that for the adoption process it is important to initially target the segment of innovators and early adopters (referring to Rogers), that is, the Fanatics. They are the first to try out new services and they need to be persuaded that the new service offers them added value. If they are convinced and accept the new service, other market segments will follow as a result of the diffusion process.

Several studies conducted throughout Europe clearly show that viewers have expressed their interest in watching television from a handheld device.[23] In particular, the quality and range of content offered by mobile TV is highly critical for user acceptance. Participants in different trials carried out by telecom operators in many European countries (Germany, France, Great Britain, Italy, and Finland) expect that all content from conventional television would also be available on the mobile, even if it is not suited for mobile viewing. The viewers want to have: good picture and sound quality; value for money; right selection of channels; single device to carry (phone); simplicity of use; service availability.

Television content specifically designed for mobile reception needs to be developed.[24] The formula "content is king" in the mobile environment has a provocative nature. It is widely recognized that an increased choice of channels and content alone would not be sufficient in determining the success or failure of mobile television. For example, consumers need sophisticated but easy tools to enable them to make the best use of the available content. What mobile television does is enable content to be presented in even more compelling formats with greater quality and convenience of access—thus, content is one of the primary drivers for the adoption of digital TV services.

At present, consumers are most likely to use mobile TV when away from home to kill time. For instance, while waiting for a bus or a train or just to keep informed when they are on the go—to get a quick news update. They will not spend hours continuously in front of their mobile phone watching TV.[25] Besides offering live broadcasting, mobile TV services must include video-on-demand services.

Another two acceptance criteria emerge from many recent studies on mobile TV: the usability of mobile TV services, strictly depending on the design of the interface and the way of navigating within the service, and the pricing of the service.

To summarize, early findings indicate that there is a certain willingness to pay for mobile TV content. For example, viewers indicated a willingness to pay between eight and twelve euros per month to access six to eight different television channels. All these data highlight a potential for mobile television development. But it is very important to underscore that these are all commercial data, provided by broadcasters and telecommunication operators. Furthermore, because this information comes from trials, it does not reflect the real behavior of users.

THE SUPPLY OF MOBILE TV IN ITALY AND ITS OPERATORS

The penetration of the mobile telecommunications service in Italy is above the Western European average at approximately 123.5 lines per 100 inhabitants at the end of 2005 (compared to a penetration rate of 109 lines per 100 inhabitants at the end of 2004). Growth rates have been substantially higher than the European average. This increase is due to innovative services and an increase in customers with multiple lines and the number of operators. At the end of 2005, 15 percent of Italian users owned a 3G handheld and three telecommunications operators—3 Italia, TIM, and Vodafone—had launched mobile TV services.

In Italy the main distribution platform for mobile TV services is UMTS (Universal Mobile Telecommunication System). All the abovementioned mobile operators offer three main types of video content:

1. TV channels in live streaming;
2. pay-per-view or video-on-demand of goals scored, video news, reality shows; video clips;
3. other content such as video MMS, games, and chats.

Without presenting what each operator provides in detail, as summarized in the table, it is interesting to observe that consumers can receive interactive services as well as the broadcast services (like those mentioned earlier).

The different trials and commercial supply of mobile TV launched in Italy show that in this country people are willing to spend time and money on mobile TV services. The mean consumption is three and a half minutes per day; and the reproduction of traditional TV content was not sufficiently appealing for the UMTS platform. On the contrary, successful contents are those created ad hoc for the platform such as the summary and preview (in three minutes) of a fiction or a reality show. Adult contents are often used during the night from 9 p.m. to 12 midnight.

As regards business models, users appear to prefer the pay-per-use model to that of subscription. Concretely, the promotional offer of 0.90 euros a day for "all you can watch" (Vodafone) has been more successful than the subscription of five euros a month for the "all inclusive" offer (TIM).

Just from this overview of mobile TV available in Italy it is clear that with mobile it is no longer possible to distinguish between point-to-point and mass communication. In fact, consumers can receive not only broadcast services, which means traditional TV programs (the expression of mass communication), but also interactive services and new forms of TV programs that are a clear manifestation of point-to-point communication. As suggested by the term, *mobile TV* embraces the two forms of communication: on one hand, mobile telephony is the expression of point-to-point communication and on the other, TV is the expression of mass communication.

With mobile TV, point-to-point communication is not distinguishable from mass communication if one of the sources of funding for mobile TV is considered: advertising. Advertising was used in the broadcasting industry for the first time by commercial broadcasters and was aimed at the masses. Nowadays, with mobile TV, there is the opportunity of having different types of advertising. This underlines that the distinction between point-to-point communication and mass communication is no longer valid. In fact, with mobile TV it is possible to transmit advertising with specific features addressed only to some users and to transmit interactive advertising. Thus, we have a new form of communication.

Table 13.1 The supply of Mobile TV services in Italy (authors' processing)

Operator	Streaming		VoD/PPV		Others	
	Content	*Price*	*Content*	*Price*	*Content*	*Price*
3 4,5 million 3G users	Sky TG 24 Fox mobile Concerti 3 3 Sport TV Cartoon Network Soap TV	Up 0,30 euros for 5-minute link	News Entertainment Concerts Football matches Soap operas Adult content …	0,09 for access to the portal 0,09 for every page visited from 0,60 to 4 for downloaded video	Backgrounds Ring tones Games Music Traffic news Ticket one cinema Chat, dating Videofiction	0,09 for access to the portal 0,09 for every page visited
TIM 0,6 million 3G users	Rai 1, Rai 2, Rai 3, Rai news24 La7 MTV Coming soon Games Network 5 music channels Isola dei Famosi	0,40 per minute promotional sale – 5 monthly	News Goals of football matches (SERIE A TIM) Calendar Videoclips Traffic news (traffic cam) Adult content	0,006 per minute (WAP) 0,04 for the download of a KB (GPRS) 0,75 or 2 for each video	Backgrounds Ring tones Games Logos …	From 2 to 4 for ring tones 1-2 per logo
vodafone 1,2 million 3G users	RAI News 24 Live! TV Music Happy Mobile Coming Soon Campioni, il sogno CNN Contents in streaming	1.5 each day "all-inclusive" Free first month Promotion: 0,90 every day from 30 September 2006	Video goals of football matches (Serie A Vodafone) News (TG5, TG1) Cartoons Fiction Adult content (Winter Olympics) …	0,19 for access to download 1-2 for each video Video goals 0,75 each match	Backgrounds Ring tones Games Approx. 500k musical pieces …	1-2 per ring tone 2,5 per game 1,5 per logo

Offering customers mobile TV services requires several different core competences that cannot easily be provided by one single company. As suggested by the term *mobile TV*, two different industries come together: the mobile industry and the broadcasting industry. Concretely, the merging of these two industries is evidence of the convergence between the telecommunication, media, and information technology sectors. Convergence is not just about technology but about services and new ways of doing business and interacting with society.

Technology convergence, of which the examples cited previously are representative, is based on the common application of digital technologies to systems and networks associated with the delivery of services. A trend towards industry convergence can be seen in alliances, mergers, and joint ventures, which build on the technical and commercial know-how of partners in order to exploit existing and new markets. And on a commercial level, convergence is a direct result of cross-fertilization between the telecommunications and broadcasting sectors, exemplified in mobile TV.

Going back to "Araldo Telefonico," it is interesting to observe that it was one of the first forms of convergence between the content industry and the telecommunication sector, as illustrated in the conclusion of the article. Although it would be useful to illustrate the value chain of mobile TV to understand better who the new operators in the development of mobile TV are, this chapter only intends to corroborate the theory that, with the advent of mobile TV, the distinction between point-to-point and multicast communication no longer seems to be valid.

CONCLUSION: DIFFERENT MEDIA, SIMILAR MENTALITIES

Araldo Telefonico and mobile TV are very different technologies for various reasons. Firstly, Araldo was a wire medium, used only in the home, while mobile TV is a wireless technology with the typical characteristics of mobility and portability. Secondly, Araldo appeared in an Italian context where telephone diffusion was restricted, while mobile TV could be considered a further mobile-phone application, a new use for one of the most successful technologies in contemporary Italy.[26] Thirdly, referring to McLuhan, with Araldo, both a new medium and a new message had to be invented: in fact, new and original editorial contents, which had never appeared before, had to be set up, as did an innovative diffusion procedure: broadcasting. On the other hand, mobile TV, even if new technology influences content format, is not a "radical invention."[27]

It is important to analyze both the successes and failures in the history of technologies: every dead medium is simply an unrealized idea of society and culture. Evidently, at the beginning of the twentieth century, receiving news and contents directly at home started to be considered a possibility, exactly as nowadays watching TV in the streets seems to have entered the "horizon of uses" of mobile phones. This use of mobile TV is not fixed and defined forever: the history of mass media teaches us that the first identified use for a product is often not the definitive one (take, for example, the phonograph and the kinetoscope). A medium is the result of the meeting between new tools and social uses,[28] and this meeting is a long and enduring sociocultural process that could even take decades.

These two technologies have two other similarities. The first is related to media convergence. Even if many scholars consider this phenomenon no older than thirty to forty years, the first example of media convergence was really Araldo Telefonico, where telecommunications and editorial contents met. And more than Araldo Telefonico, the mobile TV is the maximum expression of media convergence, characterized by the integration of different infrastructure, services, and businesses.[29] In particular, with mobile TV it is possible, thanks to the convergence between informatics and telecommunications, to watch audiovisual contents everywhere.

In particular, this paper has focused on this second similarity. While Amparo Lasen has previously discussed the continuity between Hirmondò broadcasting and 3G services,[30] we focused specifically on the fact that, for the two technologies mentioned earlier, it is impossible to distinguish between point-to-point and broadcasting. Araldo Telefonico caused a legislative crisis and drew attention to a shortcoming in definitions. Mobile TV is causing another revolution, similar to Araldo's (for this reason we have used the term *back to the future*). While throughout the twentieth century point-to-point communications (telegraphs, telephones, mobile phones) and broadcast communications (radio and TV) were identified from a legislative, technological, and managerial point of view, now, with mobile TV, these two techno-mental representations seem to lose evocative power.

The communication model adopted by both Araldo and by mobile TV cannot be defined as either point-to-point (what we could call one-to-one) or as broadcasting (one-to-mass). These two media are better characterized by the formula one-to-many: in fact, in both media, contents are received *only* by people who paid for a subscription and who own the technical receiver.

Without considering the best terminology (one-to-many or broadcasting) to use, the aim of this paper was to illustrate how two media, so different from many points of view, require a similar "mental horizon." If before the 1910s the distinction between point-to-point and broadcasting (or one-to-one and one-to-many) could not only not be understood but not even imagined, in the first decade of the twenty-first century these two concepts appear confused again. The medium of personal communication par excellence, the mobile phone, allows a use which is quite similar to traditional mass media: is it the tool-telephone *tout court* (fixed, video, or mobile) that allows this confusion?

NOTES

1. Domenico Civita, *Telefono: Estratto dall'Enciclopedia delle Arti e Industrie* (Torino: Unione Tipografico-Editrice, 1895), 6.
2. Sydney H. Aronson, "Bell's Electrical Toy," in *The Social Impact of the Telephone*, ed. Ithiel De Sola Pool (Cambridge: MIT Press, 1977), 22.
3. Ministero dei Lavori Pubblici, *Relazione statistica sui telegrafi del Regno d'Italia nell'anno 1879* (Roma: coi Tipi di Ludovico Cecchini, 1880), 36.
4. "Dappertutto telefoni," *L'illustrazione Italiana*, January 1, 1882, 86.
5. Carolyn Marvin, *When Old Technologies Were New* (New York: Oxford University Press, 1988).
6. "Il concerto telefonico," *Il Telegrafista*, 5 (1887): 127.
7. Peppino Ortoleva, "Il tempo della radio: Piccola storia del segnale orario," *Movimento Operaio e Socialista*, 2 (1986): 315–20.
8. Business manuscript "Araldo telefonico" company (1913–1914).
9. Alessandra Bartolomei and Paola Bernabei, *L'emittenza privata in Italia dal 1956 a oggi* (Torino: Rai-Eri, 1983).
10. "Araldo non costituisce conversazione telefonica," *Rivista delle Comunicazioni*, June 1917: 166.
11. "Araldo non costituisce," 166.
12. "Araldo non costituisce," 167.
13. "Araldo non costituisce," 168.
14. Peppino Ortoleva, "Il videotelefono," in *Oggetti d'uso quotidiano*, ed. M. Nacci , (Venezia: Marsilio, 1998).
15. Asa Briggs, "The Pleasure Telephone: A Charter in the Prehistory of the Media", in *The Social Impact of the Telephone*, ed. Ithiel De Sola Pool (Cambridge: MIT Press, 1977), 41–48.
16. Digitag, "Television on a Handheld Receiver," 2005, http://telin.ugent.be/~kayzlat/DVB-H/DigiTAG-DVB-H-Handbook.pdf.
17. Digitag, "Television on a Handheld Receiver."
18. Josef Trefzger, *Mobile TV-Launch in Germany—Challenges and Implications* (Cologne: Institut für Rundfunkökonomie and der Universität zu Köln, 2005).
19. Roger Silverstone and Leslie Haddon, "Design and Domestication of Information and Communication Technologies: Technical Change and Everyday Life," in *Communication by Design*, ed. Robin Mansell and Roger Silverstone (Oxford: Oxford University Press, 1996), 44–74.
20. Everett Rogers, *Diffusion of Innovations* (New York: The Free Press, 1962).
21. Frank M. Bass, "A New Product Growth Model for Consumer Durables," *Management Sciences*, 15(2) (1969): 215–27.
22. Michael Amberg and Jens Wehrmann, eds.,"Benutzerakzeptanz mobiler Dienste: Ein Erfahrungsbericht zum Compass-Akzeptanzmodell," in *Arbeitsbericht, Wirtschaftsinfromatik*, 3 (2003): 5.
23. Bmco, "The bmco Project in Berlin on Novel, Interactive and TV-like Services for Mobile and Portable Devices Has Been Completed Successfully," October, 13, 2004, .http://www.bmco-berlin.com/docs/bmco_Pressrelease_english_041013.pdf; Nokia, "Mobile TV Broadcasting," 2005, http://www.mobiletv.nokia.com/
24. Visiongain, "Mobile TV: Market Analysis and Forecasts 2004–2009," 2004, 156–57.
25. Caj Södergard, "Mobile Television—Technology and User Experiences" (2003, http://www.milab.dk/dokumentation/public/Rapporter%20om%20mobiltelefoner,%20IT%20etc/VTT%20-%20Mobile%20Television%20Technology%20and%20User%20Experiences.pdf), 167.

26. Gabriele Balbi, "Dappertutto telefonini. Per una storia sociale della telefonia mobile in Italia," *Intersezioni* (forthcoming).

27. Thomas P. Hughes, "The Evolution of Large Technical Systems," in *The Social Construction of Technological Systems*, ed. Wiebe E. Bijker, Thomas P. Hughes, and Trevor Pinch (Cambridge and London: MIT Press, 1989).

28. Jacques Perriault, *La logique de l'usage: Essai sur les machines à communiquer*, (Imprimé en France: Flammarion, 1989); David E. Nye, *Technology Matters: Questions to Live With* (Cambridge and London: MIT Press, 2006).

29. Benedetta Prario, *Le trasformazioni dell'impresa televisiva verso l'era digitale* (Bern: Peter Lang, 2005); European Commission, *Green Paper on the Convergence of the Telecommunications, Media and Information Technology Sectors, and the Implications for Regulations* (Brussels: EC, 1997).

30. Amparo Lasen, "History Repeating? A Comparison of the Launch and Uses of Fixed and Mobile Phones," In *Mobile World: Past, Present and Future*, ed. Lynne Hamill and Amparo Lasen (London: Springer, 2005), 31–32.

14 Net_Dérive

Conceiving and Producing a Locative Media Artwork

Atau Tanaka and Petra Gemeinboeck

INTRODUCTION

The rapid uptake of mobile telephony, the high-bandwidth network access afforded by third-generation (3G) networks, and the increasingly powerful multimedia capabilities of modern mobile handsets have created a potential for new forms of creative cultural works to be conceived for, and deployed exclusively on, mobile devices and wireless infrastructures. However, commercial offerings in the area of mobile media (available at the time of writing) indicate a tendency to simply extend traditional media into the mobile sphere. Typical propositions include television on the mobile phone and downloading music to the mobile. What are the forces and constraints that limit present-day usage on the mobile to simply parallel what we do in the living room or on a stationary computer? Instead of replicating traditional media in a portable package, can we not look at the mobile phone and high-bandwidth mobile networks as an artistic canvas on which to create entirely new forms of art? We broach this question with a focus on mobile music, and present Net_Dérive, a multiuser mobile artwork.

We first introduce the field of mobile music and identify the characteristic specific to music making in the mobile sphere. We next present conceptual issues in composing a scenario that leads to an abstract shared narrative. We describe the implementation of these concepts in a working hardware/software framework and discuss the issues involved in exhibiting the work. We will see from a musical perspective how participative, flexible content forms can be created.

MOBILE MUSIC

There are increasingly sophisticated commercial offerings in the marketplace that combine the functions of a personal music player with that of a mobile phone. These products often co-opt the term *mobile music*.[1] While they indicate the potential convergence of functionality and point out interesting trends, they could benefit from research of the sort presented in this book—of user-centered design, cultural theory, or in the case of this chapter, investigations of new media content composition strategies.

Other areas where commercial and grassroots developments have brought about new audio formats include ring-tones and podcasts. Ring-tones are one of the main ways of personalizing a mobile. The music industry, as much as it resisted the shift to downloadable music, has embraced ring-tones as a significant revenue source. This represents not a new market but represents the shift of purpose of music as an *accessory* for the device, to heighten its personal uniqueness.[2]

Apple's iPod/iTunes system brought with it a form of audio distribution, the podcast, which allows RSS feeds of audio content to be heard on a personal music player. It maintains the radio model of broadcast, transplanted to a download-and-synchronize mode of usage. It should be noted that podcasts are heard in deferred time. They are not live, but are produced, recorded, stored, and downloaded as static audio content. In this sense they do not adhere to the "Anytime, Anywhere" spontaneous credo of mobility.[3]

While personal music players and cell phones share many qualities—they are portable devices; they are audio devices; they are highly personal devices—in the end they each serve very different social functions. Here we discuss the conceptual foundations that led to the production of Net_Dérive, an artists' project that seeks to exploit synergies creatively and identify ways to couple communication channels directly to musical form.

CREATIVE POTENTIAL OF THE MEDIUM

To better understand the nonobvious nature of merging the personal music player and cell phone, we look at social functions of these devices. More specifically, we identify these distinctions and seek out ways in which to bridge these differences to imagine what new forms of music and multimedia could emerge from this combination of technologies that would not have arisen from nonmobile contexts.

The Walkman is a cultural icon that evokes a sense of mobility and ubiquity of one's personal music. This gave rise in generalized form to the personal music player, which often allow a user's entire music collection to be stored and transported in compressed data formats such as MP3. Du Gay[4] and Hosokawa[5] discuss the cultural impact of the Walkman, du Gay using its introduction as a kind of technology artifact in a case study for a cultural studies analysis. It can be thought of as a kind of a posteriori form of Gaver's "cultural probe"[6] or Hutchinson et al.'s "technology probe."[7]

A mobile telephone also evokes a sense of mobility, but from the perspective of connectivity with a community of friends.[8] There is a natural tendency to map the cultural analysis done with the Walkman to the mobile phone, with the desire heightened when dealing with questions of music on these portable devices. Goggin[9] and Hemment[10] draw respectively on du Gay and Hosokawa to tease out this connection. While he recognizes the attractiveness of the comparison, Goggin notes that the mobile phone, unlike the Walkman, "is not a stand-alone technology," and that "the complexity of the commercial, technical, and regulatory characteristics of the cell phone are not brought together by one company, brand, or technology in the same way as the Walkman." Hemment draws upon Hosokawa's contextualizing of Walkman use as *doubling* media space and physical space in the way that contemporary locative media does, however, concludes by remarking that the "Walkman user serves as a metonym of the atomistic individualism of the 1980s" while "the mobile phone could be taken to stand for the connectedness of contemporary global societies. . . ."

A fundamental difference between a personal music player and a mobile phone, then, is the private, nearly isolated experience of headphone music listening with the anywhere-anytime communications of mobile telephony. Users in interviews report that they use personal music players to block out the external environment, to create their own private sphere.[11] Meanwhile, users of mobile telephones report a desire to be in continuous touch with the outside world. These two desires, both grounded in the desire to control one's immediate environment, seem diametrically opposed. The distinction, however, is not so clear cut, as music, whether listened to in isolation or not, helps a listener forge her public image to be displayed and projected.[12] And while mobile-phone users want to be reachable at any time, they also want to control accessibility so as not to be bothered at inopportune moments or at times when they desire privacy.[13] Reconciling these personal and public spheres becomes a central concern in the design of a mobile system and is crucial in the consideration of a networked interpersonal music system.

TURNING THE CITY INTO A CANVAS

The use of mobile, networked, location-aware computing devices to involve participants in mapping processes, social networking, or artistic interventions is often associated with the field of artistic practice called "locative media." Transforming geographical space into a canvas to be inscribed with personal narratives, the field offers a powerful instrument for communities of participants to coauthor and shape their environment and to collectively organize and share subjective experiences.

Locative-media practices operate in the paradoxical space[14] between two antagonistic forces: the bottom-up approaches of collaborative spaces and collective interventions and the top-down strategies of centralized power and remote control. The reliance on positional precision and location-based context also critically link location-based practices to the arena of cartography and its dominant practices of mapping. Mapping is a cultural, political, and epistemological activity and has always been a powerful instrument for masking difference, making borders, and producing coherent identities.[15]

Looking critically at the ambivalences involved in deploying these technologies bears the potential of opening up a *third space*;[16] a space for intervention, in which these power relations can be investigated and negotiated.

Electronic networks and mobile artistic interventions have transformed the notion of the everyday, the public, and the potential of social relations therein.[17] Mobile social software puts social networks into the mobile sphere, with projects like *Dodgeball* allowing friends and friends of friends to meet serendipitously as they move about a city.[18] Interaction with the immediate surroundings is added in projects tagging physical space, such as *Socialight*, *PlaceLab*, *PlaceEngine*.[19] *Sonic City* draws upon notions from ubiquitous computing and wearable computing to create a portable music system that responds via sensors to changes in the ambient environment as the user walks through an urban environment.[20] Looking at the wider scope of public authoring, the research group Proboscis developed a mobile software platform for annotating geographic places with text, images, and sounds. Their projects, *Urban Tapestries* and *Social Tapestries,* demonstrate new social and cultural benefits of new mobile technologies through playfully sharing local knowledge.[21]

THE COMPOSITIONAL MATERIALS OF MOBILITY

Music has always brought artistic insight to popular culture, where musicians remark on societal and technological change through new musical forms. Here we give some examples where developments in recorded media have given rise to new musical genres. The 45-rpm vinyl record holds only four minutes of music per side, helping define the rock 'n' roll single. With the 33 rpm and its twenty minutes per side, artists conceived the concept album. When the total time went up to seventy-four minutes with the CD, musicians stopped trying to fill the disc with music and instead made *unexpected use of the medium*, taking advantage of the Red Book CD specification to play artistically with pre-gap or simply leaving a long gap of silence at what seemed like the end, putting a surprise track long after the listener thought the album was finished. The characteristics of new recording media, then, have directly driven the conception of new musical styles.

It seems natural, then, that artists would seize the possibilities afforded by network distribution of music and formats like MP3 to once again create new forms of music. However, while the Internet has transformed music distribution and dramatically enhanced access to less known music, it has yet to have a direct impact on musical style.

The term *idiomatic writing* is used to describe the process of composing music that takes into consideration the characteristics of an instrument. If we apply the principle of idiomatic writing to mobile devices and infrastructures, the challenge becomes to find the distinguishing characteristics of the medium that might lead to musical qualities specific to that medium.

In the case of mobile telephony, qualities that we identify as defining the medium include:

- dynamic location
- multiuser contexts
- *in vivo* situation in the urban environment
- bidirectional dynamic (audiovisual input as well as output)

In the point of view put forth here, these are the qualities of the artistic canvas proposed by mobile telephony. The challenge that confronts the artist is how to create music that responds in a profound way to these locative, social, interactive dynamics.

The use of interactive, participatory, and location-sensitive technologies poses an interesting set of compositional, aesthetic, and technological challenges. First, the materials with which one must compose are largely derived from unpredictable sources. Second, the incoming data must control visual and sonic instruments that drive the evolution of the piece in perceivable ways. Third, these audiovisual real-time instruments must respect and support the conceptual motivations behind the mobile-culture dynamics. Finally, working with mobile technologies is simultaneously a glance into the future but also a reminder of the past; many amenities, functionalities, and power of modern computing technology on which we depend were only beginning to be available on mobile platforms at the time of writing.

The goal in the present work was to map the locative data of multiple users to a single musical flow. The incoming data that we sought to transform into musical information included geographic location of the users, the sounds in their immediate environment, as well as the visual field surrounding the user. Sounds picked up in the environment as well as images captured by the on-device camera become the raw materials to be sculpted. From these raw materials, advanced signal-processing techniques are utilized to build software with the articulate potential of musical instruments. Analysis-resynthesis techniques are applied to these raw materials to create variations and augmentations of the source sound that still retain their connection to the original. For example, street sounds from the neighborhood are spectrally analyzed in real time. The analysis results are then applied to other audio instruments in order to harmonically tune them to the environmental sounds. Cartography and data visualization play important roles. Collective parameters like the area covered by all of the mobile participants or their average distance from the gallery affect the overall diffusion, presence, periodicity, and reverberation of the resulting music. Simultaneously, each mobile player is also sonified independently by a number of dedicated instruments; individual parameters like each mobile participant's level of activity, speed, or proximity to the other players are used to control these dedicated instruments.

NET_DÉRIVE

The artwork Net_Dérive was realized to examine the compositional, aesthetic, and technological questions posed earlier. It is an installation piece extending beyond the confines of a gallery, to include the urban environment. Deployed on portable, networked, location-aware computing devices, the goal was to apply the notion of idiomatic composition to create a kind of musical instrument, thinking of the city-as-instrument.

Net_Dérive was presented in October 2006 at the gallery Maison Rouge in Paris. A geographic zone in the Bastille quarter surrounding the gallery became the active zone in which participants' GPS coordinates fed the audiovisual generation system. Participants' meanderings in this neighborhood became data to go through abstract visualization and sonification techniques. Participants pick up a mobile device at the gallery site. The "device" is a spandex scarf inspired by the field of wearable computing, inside which are two mobile phones and a GPS module (Figure 14.1). They are given a "mission" to carry out in the surrounding neighborhood. The screen of the phone serves as a graphical display visible at one end of the scarf. A pair of headphones is connected to this phone. The GPS unit reports geographical location to the server. The second phone automatically takes a series of photographs using its built-in camera. These images are auto-uploaded over the mobile network to the server.

Figure 14.1   The locative device consists of two mobile phones and a GPS module inside a custom made spandex scarf.

A purpose-built live multimedia program generates an audiovisual stream based on this information, and is fed back live to each mobile client. GPS data reporting the positions of the three users drive the rhythm of radarlike blips, creating a shifting polyrhythm of tones, each blip tone representing a user. Voice instructions suggesting paths to follow or turns to make are triggered by certain latitude/longitude combinations, and heard by the user, abstracted in a musical fashion. As the user chooses to heed or ignore these instructions, a trace of his path is carved out in the city. Finally there was an audio upstream from each mobile serving as a roaming live microphone with each user. These sounds of the streets were cut up, looped, processed, and mixed to be played under the blips and voice commands. This applied notions of *musique concrète* to compose with and activate a listening of real-world sound objects. With this "concretized" mix streamed to the mobile back in real time, this process became live and locative, giving the listener an abstracted soundscape to color her perception of her immediate surroundings.

The simultaneity, history, and memory of the various users' paths and uploaded images shape a series of abstract visualizations and sonifications of the incoming media and data that we consider a collective narrative and that is summed together and projected in the main gallery space. There is a feedback created as the users' actions generate the collective narrative that in turn direct them.

METAPHORS FOR CREATING AN ABSTRACT NARRATIVE

The approaches described earlier, including the musical mapping of incoming data, the paradoxical space of locative practices, and the technical framework, then shaped the task to compose with the materials at hand and to develop conceptual underpinnings to tie them together. This sought to place the technologies called upon not as neutral but as charged with cultural association. Preexisting conceptions among the users of technology, urbanity, and the mobile telephone as societal object forcibly color their perception of the work. By playing on these preconceived notions, we could use them as a springboard to invite the users to extend their sense of the culturality that mobile technology affords. In order to connect to existing cultural associations and extend these towards an abstract conceptual space, we identified several metaphors that provided familiar points of departure.

With the technical functionality and usage predicated on location tracking, audiovisual media upload, and urban navigation, we put in place metaphors of *archaeology, surveillance society*, and *air-traffic control*.

Surveillance was the most straightforward association to draw upon the collective conscience with regard to location tracking. When performing geographic localization, we are following the movements of the user in an invisible way. It recalls notions of a Big Brother control society where the authorities have continuous awareness of a citizen's activities.[22] From this perspective, the act of using mobile phones to track location needs to be looked at critically, and becomes an act of appropriation to be used in creative purposes such as the narrative instruments we created that observe and interpret the player's movements by means of abstract visualization/sonification.

We sought to use the metaphor of surveillance in two ways, direct and in inversion. The direct interpretation of surveillance rests on the simple notion that the users are being monitored without their explicit consent. At the same time, inasmuch as surveillance technologies are today a reality, societal concerns of a control society seem lessened. Why is it that the general public does not seem concerned with questions of invasion of privacy that advanced communication technologies potentially represent? One explanation is that the omnipresence of consumer devices has created a revolution in *grassroots journalism*. Beginning with the Rodney King video in 1991 denouncing police brutality,[23] amateur video has often been the first camera on the spot, be it for the tsunami in Southeast Asia or the crash of the Concorde.[24] The proliferation of blogs and decentralized information dissemination perhaps gives a sense of empowerment to the public, from which they are perfectly content to give up some privacy in exchange for empowerment. The end user, then, is today as much in power to report on events in the world as the establishment. The question of balancing of power, or tension between powers, respectively, and its potential inversion of observer and observed create the second dynamic we wished to put in place in the work.

Air-traffic control takes location tracking and adds elements of multiuser coordination and commands. In our work, this adds a dimension of control dynamic in whether participants would heed commands from a "mission command central" or not. The conceptual layer has also directly informed the audiovisual metaphor for displaying the participants' location, which we will describe in more detail later (Figure 14.2).

Finally, archaeology brings a totally different kind of metaphor, adding the dimension of memory and historical accumulation over time. Archaeology, a discipline that is commonly referred to as studying cultures of the past, provides critical concepts for mapping people and objects in space and time.[25] We explored the archaeological metaphor as a way to create a dynamic through which the users' paths are inscribed as traces in the memory of the system that could be excavated later in time by other users. We wished to transpose this discipline from a study of the past to a way of looking at the present, the recent past, and the instant. The permanent movement in urban environments creates daily strata not so much to be excavated but to be swept over by street cleaners. The tracing technologies we put to use allow us instantaneously to look at these quotidian microhistories. The act of walking carves grooves into the city asphalt and expresses the notion of depth under the city streets; an archaeology of the instant along the urban fabric's vertical axis (Figure 14.3).

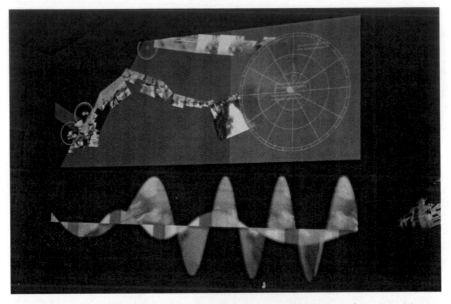

Figure 14.2   A radar-like visualization shows the participants' relative position as they unwind and intertwine their image trails.

Figure 14.3   Participants carve trails into the city skin and excavate traces that have been left by previous dwellers.

With these metaphors we sought to create a technology platform to connect sound and abstract visuals to conceptual notions of urbanity. The contemporary city is no longer only a significant geographical entity but a complex fabric of interlaced physical and virtual networks. Forming a fluid terrain, these networks link and constantly relink urban spaces. The dynamic connections not only manifest in electronic flows of information but also in the form of subjective geographies—memories, identities, and other forms of belonging. *Net_Dérive* explores this multiplicity by looking at the city through the "fictitious thickness"[26] of another—invisible—city, the virtual spaces spanned by mobile technologies.

THE DÉRIVE

The audiovisual language of our work, its sonificiations, visualizations, and spatializations of "grooving," "excavating," and "shifting" allude to the idea of the Situationist dérive (drifting). Reinvoking forms of urban intervention, such as the psychogeographic dérive, the work's collectively constructed temporary spaces encourage encounters from outside of our contained, routine driven realities. The interventionist walking practice of the dérive developed by the Situationist International was in Guy Debord's words "a technique of rapid passage through varied ambiances."[27] It involved playful-constructive behavior and, different from the classic notions of stroll,[28] aimed at interrupting the everyday.

The discourse of locative media recontextualizes the narrative of the drift that always remained open, contingent, and shifting.[29] Our work reinterprets the concept of the dérive through a dynamic playground that directs the participants' actions and that, in turn, evolves and changes in response to them. The evolution of the playground is driven by serendipitous encounters of the participants drifting through the urban environment. Inspired by the Situationist concept of "behavioral disorientation,"[30] the participants' drift is accompanied by vocal instructions, for example, "take a left here and go east for two minutes." Meanwhile, their encounters are tracked, mapped, and put in context with other encounters, with the objective to develop alternative connections. Mapping as a cultural activity here serves as a force of diversification rather than homogenization.

The playground of *Net_Dérive* creates a tension field between actors (participants) in the urban environment and a surveying gallery space. It produces an archaeology of the instant, enabling the surface of the city to be carved and inscribed while these grooves are instantaneously uncovered, contextualized, and displayed. As participants drift through the neighborhood, carving the grooves, the gallery functions as a kind of control center that processes and displays the sonic and visual excavations. Feedback of this central process is sent back to the mobile phones, making the drifters aware of the others' location and revealing the traces they captured from their environs. These traces in form of automatically recorded sounds and images become the building blocks of the excavation site. The recontextualization of excavated artifacts composes a sonic and visual scape of the participants' collective drifting. This composition, in a way, detaches the paths from the "drifter" and performs its own capability to "drift." Present and past merge and intertwine as the mobile participants "excavate" earlier traces left by previous participants.

The city's "memory" thus materializes as a contact zone at which the dichotomies of present–past and real–virtual meet and interlace, and eventually dissolve.[31] Grooving into the layers of recent history, *dérive* is not only horizontal but also vertical.

PARTING THOUGHTS

We are living in an increasingly mobile culture. As society has the means to be increasingly mobile, what are the real issues in creating a culture of mobility? Commercial offerings simply attempt to transpose existing media, such as television, to portable devices. This passive consumerism is paralleled by current commercial developments that primarily utilize location-sensing features for location-based advertising and services. It appears as if our current mobile media culture not only reinvokes Guy Debord's urban critique but also mirrors the source of this critique, namely, the spectacle that turns the individual into a passive consumer.

For the technology journalist Howard Rheingold, today's most important critical question is whether "entire populations of city-dwellers [will] create, use, and exchange information and media associated with geographic locations," or whether we will "be passive consumers of pre-packaged content fabricated by a few dozen synthetic superstars."[31] We could already witness unexpected usage emerging in a grassroots fashion to have cultural impact, for example, the explosion of SMS leading to new linguistic vernaculars and spellings. The project presented here offers a glimpse at how we can leverage these forces to create deeper cultural experiences for a mobile society. With Net_Dérive we created a musical locative work.

Market-driven products rarely fulfill the true creative potential of a medium. The commercial offerings in mobile music are no exception. We have the possibility, from the point of view of fundamental research, to take an alternative angle to tackle these problems. By building systems as artworks, we gain insight often overlooked in the rush of product development. Injecting creativity into the research process allows us to better understand the creative potential of mobile media.

Creating working prototypes for such ideas is an act of system building. We created an architecture of itinerant devices connected over wireless infrastructures, utilizing network-level services such as localization, to feed a dynamic audiovisual content-generation process. With all this, the primacy still resided in thinking about what the resulting music sounded like and what the ultimate experience would be for the listener. Composing for such an environment is to look at the system as a kind of instrument, a musical instrument that is not an acoustic network of vibrating elements and air impulses but a human/technological network of entities and elements coming into musical interaction.

NOTES

1. Vasilios Koutsiouris, Pavlos Vlachos, and Adam Vrechopoulos, "Developing and Evaluating Mobile Entertainment Applications: The Case of the Music Industry," ed. Matthias Rauterberg, *Entertainment Computing–ICEC 2004, Lecture Notes in Computer Science*, 3166 (Berlin: Springer, 2004); Per Andersson and Christopher Rosenqvist, "Mobile Music, Customer Value, and Changing Market Needs," *The International Journal on Media Management*, 8 (2006): 92–103.
2. Tomoyuki Okada, "Youth Culture and the Shaping of Japanese Mobile Media: Personalization and the *Keitai* Internet as Multimedia," in *Personal, Portable, Pedestrian: Mobile Phones in Japanese Life*, ed. Mizuko Ito, Daisuke Okabe, and Misa Matsuda (Cambridge: MIT Press, 2005), 41–60.
3. Mark Perry, Kenton O'Hara, Abigail Sellen, Barry Brown, and Richard Harper, "Dealing with Mobility: Understanding Access Anytime, Anywhere," *ACM Transactions on Computer-Human Interaction*, 8 (2001): 323–47.
4. Paul Du Gay, *Doing Cultural Studies: The Story of the Sony Walkman* (Berkshire, UK: Open University Press, 1997).
5. Shuhei Hosokawa, "The Walkman Effect," in *Popular Music Vol. 4: Performers and Audiences*, ed. R. Middleton and D. Horn (Cambridge: Cambridge University Press, 1984), 165–80.
6. Bill Gaver, Tony Dunne, and Elena Pacenti, "Design: Cultural Probes," *Interactions*, 6(1) (1999): 21–29.
7. Hilary Hutchinson, Wendy Mackay, Bo Westerlund, Benjamin B. Bederson, Allison Druin, Catherine Plaisant, Michel Beaudouin-Lafon, Stéphane Conversy, Helen Evans, Heiko Hansen, Nicolas Roussel, and Björn Eiderbäck, "Technology Probes: Inspiring Design for and with Families," in *Proceedings of the SIGCHI Conference on Human Factors in Computing Systems*, Ft. Lauderdale, Florida, 2003.
8. Sadie Plant, "On the Mobile: The Effects of Mobiletelephones on Social and Individual Life" (Motorola, 2001 http://www.motorola.com/mot/doc/0/234_MotDoc.pdf).
9. Gerard Goggin, *Cell Phone Culture* (London: Routledge, 2006).
10. Drew Hemment, "The Mobile Effect," *Convergence* 11 (2005): 32–39.
11. Michael Bull, *Sounding Out the City: Personal Stereos and the Management of Everyday Life* (Oxford: Berg, 2000).
12. Tia DeNora, "Music and Emotion in Real Time," in *Consuming Music Together: Social and Collaborative Aspects of Music Consumption Technologies*, Computer Supported Cooperative Work, vol. 35, ed. Barry Brown and Kenton O'Hara (Dordrecht, Netherlands: Springer, 2006), 19–33.
13. Christine Satchell, "The Swarm: Facilitating Fluidity and Control in Young People's Use of Mobile Phones," in *Proceedings of OzCHI 2003: New Directions in Interaction, Information Environments, Media and Technology*, 26–28, November 2003 (Brisbane: Information Environments Program, University of Queensland, 2003).
14. Drew Hemment, "Locative Dystopia 2," in *TCM Locative Reader*, 2004, http://www.locative.net/tcmreader/index.php?locarts;hemment-dystopia.
15. Irit Rogoff, *Terra Infirma: Geography's Visual Culture* (London: Routledge, 2000).
16. Homi K. Bhabha, *The Location of Culture* (London: Routledge, 1994).
17. Marc Tuters, "The Locative Commons: Situating Location-Based Media in Urban Public Space," Futuresonic (2004), http://www.futuresonic.com/futuresonic/pdf/Locative_Commons.pdf.

18. Ian Smith, "Social-Mobile Applications," *IEEE Computer* 38(4) (2005), 84–85.
19. Anthony LaMarca et al., "Place Lab: Device Positioning Using Radio Beacons in the Wild," *Pervasive Computing, Lecture Notes in Computer Science*, 3468 (Berlin: Springer, 2005).
20. Layla Gaye, Ramia Mazé, and Lars-Erik Holmquist, "Sonic City: The Urban Environment as a Musical Interface," in *Proceedings of NIME03* (2003), 109–15.
21. Giles Lane, "SOCIAL TAPESTRIES: Public Authoring and Civil Society," in *Proboscis Cultural Snapshots*, 9 (2004), http://research.urbantapestries.net/socialtapestries.html.
22. George Orwell, *Nineteen Eighty-Four: A Novel* (London: Secker & Warburg, 1949).
23. Robert Gooding-Williams, *Reading Rodney King/Reading Urban Uprising* (London: Routledge, 1993).
24. BBC, "Concorde: What Went Wrong?" *BBC News*, September 5, 2000, http://news.bbc.co.uk/2/hi/europe/851864.stm.
25. Anne Galloway and Matt Ward, "Locative Media as Socialising and Spatialising Practice: Learning from Archaeology," *Leonardo Electronic Almanac*, 14(3–4) (2006), http://leoalmanac.org/journal/Vol_14/lea_v14_n03–04/gallowayward.asp.
26. Italo Calvino, *Invisible Cities*, trans. William Weaver (London: Vintage, 1974).
27. Guy Debord, "Theory of the Derive," in *Situationist International Anthology*, ed. Ken Knabb (Berkeley, CA: Bureau of Public Secrets, 1981), 50–56.
28. Simon Sadler, *The Situationist City* (Cambridge, MA: MIT Press, 1999).
29. Petra Gemeinboeck, Atau Tanaka, and Andy Dong, "Instant Archaeologies: Digital Lenses to Probe and to Perforate the Urban Fabric," in *Proceedings of ACM Multimedia 2006* (New York: ACM Press, 2006), 279–86.
30. Guy Debord, *The Society of the Spectacle*, trans. Donald Nicholson-Smith (New York: Zone, 1995).
31. Howard Rheingold, "Urban Infomatics Breakout," *The Feature: It's All about the Mobile Internet* (2004), http://www.thefeaturearchives.com/topic/Culture/Urban_Infomatics_Breakout.html.

15 Mobile News in Chinese Newspaper Groups

A Case Study of Yunnan Daily Press Group

Liu Cheng and Axel Bruns

> Today the cell phone has become so ubiquitous that its wonders to behold are commonplace, an astonishing part of everyday life.[1]

Perhaps the most rapid developments in adoption and usage patterns for mobile or cell phones can today be observed in Southeast Asia, where the Japanese and South Korean markets for mobile telephony are already mature, and where the Chinese market is set to overtake them very quickly in terms of quantity but perhaps also in terms of technological sophistication. It is therefore important to observe the developing uses of mobile-phone-based communication and to examine the strengths and weaknesses of the Chinese approach to mobile technologies, across a variety of social, commercial, and official domains. As approaches to mobile telephony mature across these domains, the technology itself is gradually finding its place and role in Chinese society—a role which may vary significantly from that in other societies. This social construction of technology is necessarily an ongoing, never-complete process negotiated between everyday users, corporate service providers, and government regulators.

In this context, it is also necessary to recognize the vast differences which exist between specific Chinese regions and localities. Uses of mobile telephony in key economic and administrative centers such as Shanghai or Beijing may be significantly different from those in less prominent and more remote areas, and a research focus only on the outstanding examples of the booming Chinese economy may well distort the overall picture. At the same time, Chinese government policy for the bulk of the country, outside of the special administrative and economic areas, is relatively uniform, so that a case study of developments outside of these special areas can be seen to offer insights which may be understood as representative for much of the rest of China.

This chapter, therefore, is based chiefly on developments in Yunnan Province, and here focuses especially on the gradual development of mobile-phone-based news services. It specifically builds on the expertise of its primary author, Liu Cheng, who has been in charge of the operation of cell phone, SMS, and Wireless Application Protocol (WAP) news services within Yunnan Daily Press

Group since 2002, as part of his role as vice-director of the information and network center of Yunnan Daily Press. Compared with its counterparts, for example, in Beijing, Shanghai, or Guangzhou, who began to utilize new and mobile media forms at an earlier stage, Yunnan Daily Press Group is still a developing newspaper group. Its approach is informed to some extent by these early adopters' experiences, as well as by the efforts of other national and regional newspaper groups around China—for example, Xinhua News Agency and *People's Daily* in Beijing, Jie Fan Daily Press Group and Shanghai Media Group in Shanghai, Nan Fang Media Group and Guang Zhou Newspaper Group in Guang Dong province. At the same time, it has also been important to translate the experiences of newspaper groups elsewhere in China to the specific needs of the Yunnan audience.

Overall, the Yunnan experience demonstrates some important trends relating to mobile-phone-based news access in China. For Yunnan Daily Press, mobile newspaper services have now become the most important area of its new media operations. At the end of 2007, the number of subscribers to its recently introduced MMS-based mobile newspaper has reached 100,000 users, with some 50,000 still accessing an older, SMS-based service; additionally, the market leader in mobile telephony, China Mobile, is going to promote the mobile newspaper of Yunnan Daily Press around the country. This makes Yunnan Daily Press an even more relevant example for mobile telephony developments in China, beyond Yunnan Province itself.

A BRIEF HISTORY OF RECENT DEVELOPMENTS IN THE CHINESE SMS MARKET

In China, the mobile phone is known as the fifth media form, closely following newspapers, broadcast, TV, and the Internet. There are two main mobile telecommunications companies in this field: China Mobile Communications Corporation (or "China Mobile") and China Unicom Limited ("China Unicom"); together, they control a large part of the private-use mobile telephony market. This market has shown impressive growth in terms of numbers of subscribers over the past six years: statistics from the Ministry of Information Industry of the PRC show that the number of subscribers rose by some 1,200 percent in the eight years from 1999 to November 2007 (see Figure 15.1).

British company Vodafone sent the world's first mobile phone short message from a PC to a mobile phone on its GSM network as early as 1992, but prior to the introduction of attractive messaging cost plans, growth in SMS services remained slow. Their development in China was further hampered by language difficulties. Tianjin launched the first Chinese-language mobile-phone short-message service in China in 1997; before this, mobile phones in China only offered English (or Latin script) short-message functions. SMS messaging and SMS-based information services began spreading rapidly around the world after 1997, but operations in China really only

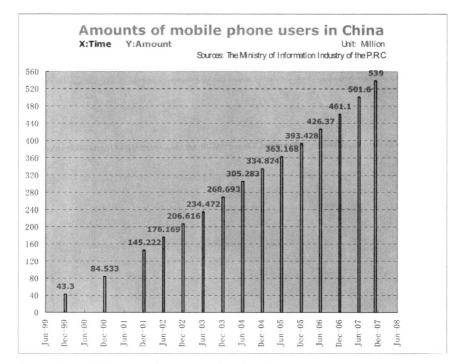

Figure 15.1   Chinese mobile phone subscribers.[2]

started when China Mobile launched the "Monternet Plan" in November 2000. It was the first mobile value-added service available in China. Monternet (mobile + Internet) acts as a bridge between mobile devices and Internet-based and other digital information services. To make use of the new service, mobile communication companies authorized a range of service providers (SPs) to deliver value-added services content by way of SMS to mobile-phone users. The monthly package fee for such services is around two to eight yuan (A\$0.333–1.333) in most cases. Users pay such fees to their mobile communication companies every month, and during that month receive about 30–100 messages.

As a result of such new services, according to statistics from the Ministry of Information Industry of the PRC, the SMS usage volume increased from one billion messages in 2000 to some 570 times that amount in 2007 (see Figure 15.2).

In 2002, the mobile-phone short-message industry was even said to have saved the Chinese Internet. The three biggest portal Web sites in China were in danger of being delisted from the NASDAQ. However, that year one of them, Netease, was the first to achieve profitability by relying on network games and short-message services, and the stock market success of Netease became the engine for other China-related concept stocks in the NASDAQ.

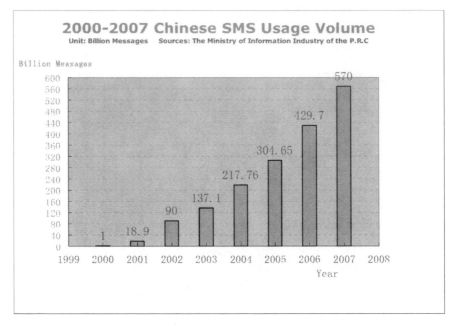

Figure 15.2   SMS usage rates.[3]

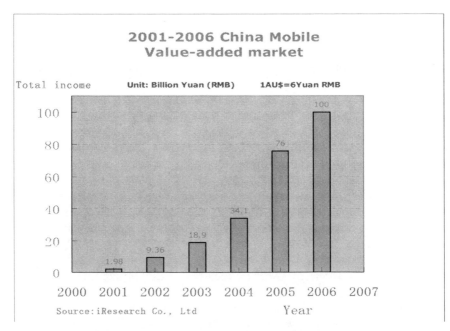

Figure 15.3   Chinese value-added SMS market growth.

Shortly after, Sina, Sohu, Tom and other Web sites were also able to generate strong profits through short-message value-added services.[4]

As a result, the Chinese value-added SMS market enjoyed five years of rapid development (see Figure 15.3), and the total size of the market reached 100 billion Yuan in 2006, and was expected to approach more than 120 billion Yuan in 2007.[5]

YUNNAN DAILY PRESS GROUP'S DEVELOPMENT OF SMS NEWS SERVICES

As they have begun to do elsewhere in the world, news publishers have the ability to make an important contribution to the provision of value-added SMS services. There are two tiers of newspaper groups in China, respectively existing at the national and at the provincial levels. The majority of newspaper groups are at the provincial level, spread across almost all of China, and their news represents the most authoritative information in China at a national and local level. Yunnan Daily Press Group ("Yndaily" for short) is the only newspaper group in the Yunnan Province of China. Yndaily combines ten newspapers, four periodicals, and one Web site (www.yndaily.com) into one group. It provides various forms of information, and especially local news, to the people of Yunnan Province.

Located in the southwest of the Chinese mainland, Yunnan is home to a population of about 45 million and is governed from the provincial capital Kunming. Its 2006 GDP was just over 400 billion yuan (US$50 billion), growing at a rate of some 11.9 percent per annum. Secondary and tertiary industries contribute the vast majority of the province's gross domestic product. In spite of such growth, however, some 2.3 million citizens continue to live in absolute poverty in the province, and some seven million live below the official poverty line. This places Yunnan at or below the national average for China.[6] Because of its comparatively ordinary performance within the Chinese context, however, Yunnan Province and its news and media organizations—chiefly, Yunnan Daily Press Group—provide a useful case study which may be more representative of the overall Chinese experience than studies of the better-known boom cities and provinces.

Yndaily developed its SMS news service in partnership with Yunnan Mobile Communication Company, which introduced its SMS services interface in the middle of 2002. Main issues for Yndaily in this process were the news content management process, the service's technical feasibility, and the development of a strategy for the new medium. While sourcing content for an SMS news service posed no problem, given Yndaily's existing resources, the company did not have technological expertise in mobile-phone service provision, and had to rely on its partner to provide this knowledge. Yndaily began to offer its SMS information service in August 2002. (Other mobile companies throughout the country also introduced SMS news services into their markets around this time.)

Of course, the quality of short-message services is key to generating user satisfaction. Yndaily sends two to five short messages to users every day. SMS is limited to a maximum of seventy Chinese characters, including numerals and punctuation, and the challenge of collecting news and remaining within the word limit is difficult. SMS news is a new project for traditional journalism, and its extreme limitations of characters per message challenge editors needing to present the headline news of the day.

PROBLEMS WITH THE MOBILE VALUE-ADDED SERVICES MARKET IN CHINA

Because this field is a comparatively new, developing market, service quality has varied from provider to provider, and initially there was a pronounced lack of management experience and governance rules for service providers (SPs). Many SPs initially exploited an absence of clear rules and regulations, to the disadvantage of their consumers. A large number of complaints from SMS consumers were voiced in 2004 and 2005, and as a consequence the authorities utilized administrative and technological means to restrict the actions of SPs. This action meant that a number of SMS content providers had to withdraw from the market gradually, but in a follow-on effect also affected the ability of legitimate and conscientious providers to do business effectively.

In the Chinese system, service providers are not entitled to charge fees from consumers directly: fees for information services are charged by mobile communications companies on behalf of the SP. However, most of the fee statistics are calculated by the SPs and forwarded to mobile communications companies at the end of every month; mobile communication companies then charge customers according to these fee data.

As there were some gaps in managing this new market, however, some SPs initially provided falsified statistical data and charged fees from users illegally—for example, by introducing additional fees without notice, or by duplicating charges. Some users also found it difficult to cancel their services: at times there was no response to users' requests, or the process of canceling services was overly complex. Some SPs also sent information to users directly, subscribing users and charging fees without the involvement of the mobile carrier; further, some SPs had no department or service hotline to deal with complaints; and in the worst cases, users were charged fees even when they had never received any information services.

To address such problems, China Mobile, the biggest mobile communication company in China, began to penalize those SPs that had been found guilty of fraudulent practices, and increased its penalties in order to manage the service providers. Ultimately, more than one hundred SPs were penalized by the provincial branch companies of China Mobile around the country,[7] and some well-known companies, such as Sina, Sohu, Mtone Wireless, HL95, and Tencent Inc., were placed on a black list. At the end of 2004 China Mobile started up a new information management platform known as MISC (Mobile Information Service Center). Using this platform, mobile communication companies could utilize technical means to enhance the management of SPs. All SPs were required to connect to this system; if SPs missed the deadline, they would lose their qualification as a service provider.

Through this platform, managing relations between subscribers and SPs became the responsibility of mobile communication companies. Furthermore, the only way that a relationship between subscribers and SPs could now be established was by subscribers actively sending short messages to service providers. SPs are unable and do not have a right to send any short messages to users with whom they have no existing relationship, even if the messages are provided for free or claimed to be useful. Service fee statistics for all SPs are also calculated by MISC system. In this way, the operational power moved from SPs to the mobile companies, and illegal behavior and harmful information can now be controlled and filtered efficiently.

Both with the help of such added customer protection measures, and due to the overall rapid growth in Chinese mobile-phone users, the development of the various value-added services offered by Internet portals and traditional media, which represented the first generation of service providers in the Chinese mobile communications market, has continued at high speed. In spite of the exploitative practices employed by some providers, as long as SPs provided decent value-added services, they were able to obtain increased market returns over time: customer churn due to some providers' unethical conduct was more than balanced out by the overall growth in the user population. As a result, a large number of further new SPs have continued to step into the field.

Gradually, too, the Chinese government has adjusted its industry regulations in an attempt to stamp out exploitative practices. Table 15.1 shows the relevant rules and regulations which have been implemented since the problems were first exposed. The first rules and regulations were enacted in September 2000; in 2004 and 2005, further regulations were introduced when additional problems emerged. (Other regulation addresses cultural, social, and political requirements for communication services, but cannot be addressed in this chapter for reasons of space.) From September 2005, the Ministry of the Information Industry has required Chinese telecommunication companies to exclude those SPs which continued to misbehave. China Unicom dropped 125 SPs from its network platform, and canceled cooperation with seventy-nine SPs. Another nineteen SPs received heavy fines from the China Telecommunication Company.[8]

Table 15.1   Relevant Rules and Regulations

Time	Name	Authorized departments
September 25, 2000	Regulation on Internet Information Service of the PRC	State Department of China
April 2004	Notice of relevant questions about standardizing the short-message services	The Ministry of Information Industries of the PRC
March 13, 2005	Criteria for telecommunication services	The Ministry of Information Industries of the PRC
September 25, 2005	Provisions for the administration of Internet news information services	The Press Office of State Department and the Ministry of Information Industries of the PRC

Source: The Ministry of Information Industry of the PRC.[9]

Such problems are hardly limited to China, of course—developing SMS services markets in other nations have similarly moved through early phases of uncontrolled growth and experimentation with service models, and from there towards a gradual consolidation and regulation of the industry. The experience of that early phase can have a lasting effect on consumer attitudes, however, as it substantially affects user trust in the mobile medium's available content, services, and service operators. It is therefore necessary for regulators to strike a difficult balance by adopting a level of regulatory oversight that is sufficiently strong to protect customers from unscrupulous operators and light enough to allow for the continued exploration of innovative SMS business models.[10]

EFFECTS ON YUNNAN DAILY PRESS'S MOBILE NEWS SERVICES, AND FUTURE POSSIBILITIES

Limited user trust remains a key barrier for the ongoing development of mobile value-added services in China, however. At present, the traditional means of publicizing value-added products through newspapers, TV, and broadcasting are not enough to attract consumers, especially to innovative service products. Some promotional means previously employed even by legitimate SPs have now become illegal—for example, service providers can no longer send invitational messages or provide unsolicited free trial services to users. The development of alternative, legal means for the promotion of their mobile information services is today a key issue for every SP.

Even though it never engaged in exploitative business practices itself, Yunnan Daily Press's mobile news services were affected by the trust crisis and subsequent downturn in mobile value-added service subscriber numbers. Statistics gathered by Yndaily show that the number of users subscribed to its SMS news services grew from 1,500 users in September 2002 to a top of 290,000 users in January 2005, but declined to 120,000 users in January 2006. This translates to a growth in users by some 10,000 subscribers per month from September 2002 to January 2005. (In the same period, 2002–06, the two main newspapers published by Yndaily—*Yunnan Daily* and *ChunCheng Evening*—maintained circulations of between 150,000 and 200,000 copies.)

How can mobile value-added service providers, and especially the providers of mobile news services, reinvigorate this market? A number of new opportunities are currently becoming available for Chinese mobile news providers, and like many of its regional counterparts Yndaily is currently in the process of exploring these options.

Leveraging Established Content Sources

The Yunnan Daily Press Group and similar news organizations hold the original copyright and intellectual property rights for their content. This means that existing newspaper groups have an immense advantage compared with other providers which are engaged in new media information services. Traditional media have set up their information services as a way to utilize the new media to provide value-added information services.

Participatory Journalism and Related Interactive Options

As Shayne Bowman and Chris Willis state, "participatory journalism provides media companies with the potential to develop a more loyal and trustworthy relationship with their audiences."[11] As Yndaily has discovered, from January to July 2006, those mobile-phone subscribers who participated in interactive activities (voting, commenting, feedback, responses to questions, etc.) were growing in number. In May to July 2006, Yndaily subscribers on average sent some 200,000 messages per month; this is a significant increase from an average of 50,000 messages per month during 2005—especially given that the total number of subscribers declined from 290,000 to 120,000 users during 2005 and remained at that level during 2006.

MMS Services

In June 2006, Yndaily established a new mobile-phone newspaper service which utilizes MMS (Multimedia Message Service) technology to send mobile-phone newspapers to users. Of course, MMS is no longer a new technology: it has long been used to transmit photos and similar messages. However, beyond simple photos, MMS also allows for the transmission of audio and video content, and Yndaily's new service utilizes these opportunities. There has been a relatively speedy transition of users to the new MMS service: the post-2005 subscriber base of 118,000 SMS news users in June 2006 has declined to 50,000 users at the end of 2007, while some 100,000 users now subscribe to the MMS-based mobile newspaper offered by Yunnan Daily Press.

3G Services

Technical tests of the 3G standard in China were conducted between October and December 2006,[12] and further commercial tests are now in progress. At a development and policy conference for the Chinese telecommunication industries in Beijing in March 2007, Lou Qin Jian, the vice-minister of the Ministry of the Information Industry, said that "the reform of telecommunication industry system will be deepened" and that "3G is an opportunity to optimise the competitive situation of the Chinese telecommunication market."[13]

However, its development remains restricted by some elements: 3G mobile phones generally remain expensive, as do the service fees for using MMS and WAP, and especially multimedia functions. It will take a long time for a developing country with a large population and low average income to popularize such devices and utilize the advanced functions of 3G mobile phones. Additionally, the Chinese mobile telephony market at present remains heavily skewed in favor of the major domestic provider, China Mobile, which is seen to stifle competition, innovation, and the introduction of new technologies. The government is therefore also using the introduction of 3G to effect changes to the structure of the industry; this, however, has repeatedly delayed the granting of 3G licenses.

3G does offer many new opportunities for mobile news providers—once 3G phones are seen as being multimedia and networking devices as much as they are telephony tools, new services are likely to no longer resemble the current SMS- and MMS-based news updates, which themselves remain based largely on traditional newspaper formats, but will probably be more closely aligned in style, content, and opportunities for user participation to comparable Internet-based offerings: they are likely to be more flexible, more customizable ,and more interactive news services. This contributes, as Jeremy Rifkin notes, to "a long term shift from industrial production to cultural production. More and more cutting edge commerce in the future will involve the marketing of a vast array of cultural experiences rather than of just traditional industrial-based goods and services."[14] However, what remains to be developed are interactive marketing and advertising approaches using 3G mobile media which actively attract, rather than annoy, their target audiences.

Yunnan Daily Press's mobile services strategy for the short- to mid-term future, then, is based around extending its existing services through the incorporation of new technologies as they become available—but this will also necessitate a fundamental restructuring of the company's overall organization as it exists today. In the short term, it is already evident that the basic SMS information service, with its limitations of seventy Chinese characters per news report, is increasingly insufficient and that readers are moving away from this service. The new MMS-based mobile newspaper service is already a success, but will have to be developed further. Yndaily believes that a subdivision of the service into specific topical channels will be a necessary and appropriate move within the user-centered new media environment.

Beyond such immediate developments, the Yndaily Information and Network Center completed a research report entitled "Building a New-Style Media Group in Yunnan Based on Integrated Information" at the end of 2006. This report suggests that Yndaily should see itself no longer simply as a newspaper group but as a multimedia company, combining Internet services, mobile communication, and further new media forms as they emerge in the future. Information technology is positioned by the report as an impetus and catalyst for changes to Yndaily's traditional operational approaches, production flows, and management strategies. Media operations are shifting from single-dimensional approaches to multidimensional models which combine a number of channels and technologies in the company's news products.

Mobile communication technologies already are amongst the most important tools towards achieving this goal. The existing SMS news service, and the new MMS newspaper, are two effective supplements to the print newspapers already offered by Yndaily, but may move well beyond playing a supportive role for conventional flagship publications. Further mobile magazines, mobile books, WAP, and mo-blog services are likely to be introduced over time as additional opportunities for publishing content and interacting with Yndaily customers and users. Ultimately, such services will be part of a cross-media suite of related and interconnected news and information products offered by Yunnan Daily Press Group, which integrates social networking, Weblog, mo-blog, WAP, and networked personal storage spaces in a community Web site. Users who subscribe to any one value-added service will be free use most of the other available functions—on the assumption that if Yndaily can provide further services on top of existing offerings, more users will continue to subscribe to its services. Finally, as part of these developments, the Yunnan Daily Press Group will also take over www.yunnan. cn, the official portal Web site of Yunnan Province. In developing such new products, Yndaily will continue to build on its existing brand advantages within the Yunnan context. Yndaily will introduce and operate its current and future mobile information services with this brand image in mind.

Of course, the organization will also closely track the development of the overall regulatory and legislative framework for press and new media organizations in China. The PRC's General Administration of Press and Publication issued its *Developmental Plan of Press and Publication Industry in*

2006 to 2010 in December 2006, highlighting digital publication as one of the strategic keystones which will be promoted over these five years. Books, newspapers, and periodicals as delivered through online, mobile, and digital television technologies are encouraged and supported by the government. As part of the development process, the China Digital Newspaper Lab, an open research laboratory, aims to explore and utilize technologies to support the growth of a digital newspaper industry. Eighteen of the most prominent newspaper and media groups around China were founding members of the lab, and Yndaily has now also been invited to join the organization. In particular, its research will explore the impact of broadband and wireless Internet as well as e-paper developments on current commercial models in the Chinese news industry; mobile and multimedia newspapers, as well as other electronic reading devices, are another key field of research.

The General Administration of Press and Publication is the administrative department overseeing all media groups around China and its guidelines directly affect Yndaily's operations—developments such as the China Digital Newspaper Lab therefore have a strong potential to enhance the mobile news services (and especially the mobile newspaper) already offered by Yndaily, as well as to lead to and inform the introduction of new services including the WAP news service, mobile magazines, mo-blogs, and other new offerings currently under consideration within the Yunnan Daily Press Group.

In spite of such positive and constructive developments, both within Yndaily itself and within the administrative frameworks of the PRC in general, it is likely that further regulatory changes will be necessary. As in many other nations, mobile information services—and especially mobile news services—still represent a new challenge for the authorities. In China, there is no one regulatory authority which manages the mobile market. The State Council Information Office of the PRC has responsibility for mobile news. The Bureau of Telecommunication Administration licenses service and content providers. Mobile communication companies offer service interfaces and networks to SPs and CPs—and at present, mobile companies are in a particularly strong position because of their control of this crucial bottleneck in the mobile communication chain. The development of a more comprehensive mobile communications authority would help streamline regulatory processes for mobile-news-service providers.

Ultimately, it is the mobile communication companies which offer information services to customers, in cooperation with media organizations or other SPs and CPs; they themselves are becoming major content providers in their own right. A monopolization of mobile information services by these providers does not benefit the development of the mobile medium in China, however. Therefore, mobile services regulation aimed at preventing monopoly effects will need to be implemented. Such regulation must address the relationship between mobile companies and SPs and CPs, and determine whether mobile companies can become media organizations themselves. Yunnan Daily Press Group's experience to date suggests that clear guidelines defining the respective roles for mobile communications companies, service providers, and content providers would be very useful to systematize the structure of the mobile communications industry in China, and to avoid market dominance by a small number of major players.

CONCLUSION

It is likely that Yunnan Daily Press' experience in offering mobile news services in the context of the developing regulatory and commercial frameworks for mobile telephony in China can be regarded as representative for the bulk of other regional news organizations in the country. While the market leaders (national news organizations as well as the regional presses in boom regions such as Shanghai or Beijing) may operate in a class of their own, outside of such extraordinary cases all Chinese newspaper groups—regardless of location—work under the same institutional, organizational, and administrative system. (In addition to the overall rules for establishing and operating a news organization, this also includes regulations for the censorship of news content, and procedures for the nomination of the top executives of newspaper organizations, for example.) Such news organizations do differ mainly in the local markets they address: they need to develop content and commercial strategies which address the specific demographic and socioeconomic realities of their home region. Based in a province which benefits from China's current rapid economic development, but cannot be seen as one of its leaders, Yunnan Daily Press Group can therefore be seen as a useful example for overall developments in the field of mobile-news-service provision in China.

What emerges from this case study is that, in the context of the rapid and sustained growth of the Chinese economy, there is a similarly rapid and sustained development of mobile information services, including news services, in Yunnan as well as in other Chinese provinces; such development (which has already seen a shift from relatively basic SMS headline services to increasingly more sophisticated MMS newspapers and is likely to turn its focus to interactive 3G services in the near future) continues to be hampered, however, by the regulatory frameworks and industry structures currently in place for mobile telephony in China.

Although divided into mobile communication companies (providing access to the networks), service providers (which handle the subscription and content provision processes themselves), and content providers (such as the Yunnan Daily Press Group), the industry is dominated by the mobile communication providers which ultimately control SPs' and CPs' access to networks and users. This may be addressed through a restructuring of the Chinese mobile telecommunications sector: instead of the current division of the industry, we may see an increasing convergence of such roles. Mobile communications providers could incorporate service provision functions, thereby also overcoming the continuing problems with service providers overcharging or falsely subscribing users to their services; they could also enter into direct partnerships with content providers to offer strong and clearly branded mobile content services. Such moves towards mergers and the vertical integration of mobile telephony industry sectors would be consistent with similar developments in other mature or maturing mobile markets around the world.

The problem with such industrial convergence, however, may be that innovation in the Chinese mobile services market could be subdued as a result: mergers concentrate more control amongst a small group of operators, and traditional content providers offering mobile services in concert with mainstream mobile communication companies may have limited incentive to innovate and might focus on relatively standard, noninteractive offerings rather than exploring new and interactive opportunities. By contrast, it is often the smaller and more flexible service or content providers in any new media market which develop the most innovative service offers. Broader, ongoing trends towards more interactive functionality in mobile technology might work against such limiting developments—but in order for such new mobile technologies to be used innovatively, more effective legal and corporate frameworks will also need to be established. Only if this is done will mobile media in China mature into a legitimate and well-respected media form comparable to print or broadcast media.

At the same time, the continuing overall boom of the Chinese economy may easily obscure even the existence of such structural problems for the mobile telephony industry: the relatively rapid growth of mobile phone usage in China, and the staggering statistics it generates, make it easy to overlook the fact that, at least in the field of news services, the mobile industry does not perform as well as it could. A focus on developments in provinces away from the current powerhouse performers of the Chinese economy may help uncover structural deficits underneath the façade of overall growth.

NOTES

1. Paul Levinson, *Cellphone: The Story of the World's Most Mobile Medium and How It Has Transformed Everything* (New York: Palgrave Macmillan, 2004), 1.
2. Figure created by authors from Ministry of Information Industry of the PRC (1999–2006), http://www.mii.gov.cn/col/col166/index.html.
3. Ministry of Information Industry of the PRC (1999–2006), http://www.mii.gov.cn/col/col166/index.html.
4. Xinhua News Agency, "Sina, Sohu, Netease: The Forecast of Three Portal Websites," January 23, 2003, http://news.xinhuanet.com/newsmedia/2003-01/23/content_692743.htm.
5. Figure created by authors from statistics in: iResearch, *Mobile Value-Added Report*, 2004–2007, http://www.iresearch.com.cn/html/wireless_service/more_free.html.

6. Yunnan Provincial Bureau of Statistics, "Statistical Communiqué of the 2006 Yunnan Provincial Economic and Social Development," 4 April 2007, *Yunnan Daily*, 6.
7. Yesky.com, "China Mobile Penalised Sina and 27 SPs Are on the Black List," September 16, 2004, http://www.yesky.com/97/1854597.shtml.
8. Xinhua News Agency, "China Telecommunication Company Takes Action, 19 SPs Were Penalised," November 9, 2005, http://news.xinhuanet.com/fortune/2005-11/09/content_3753820.htm.
9. Table created by authors drawing upon the following Ministry of Information Industry of the People's Republic of China sources: "Regulation on Internet Information Service of the PRC," 2005, http://www.mii.gov.cn/art/2005/12/15/art_523_1323.html; "Provisions for the Administration of Internet News Information Services," 2006, http://www.mii.gov.cn/art/2006/03/06/art_524_7500.html; "The Criterion for Telecommunication Services," 2005, http://www.mii.gov.cn/art/2005/12/17/art_524_1656.html.
10. See: Melody M. Tsang, Shu-Chun Ho, and Ting-Peng Liang, "Consumer Attitudes toward Mobile Advertising: An Empirical Study," *International Journal of Electronic Commerce*, 8(3) (2004): 65–78; Amy Carroll, Stuart J. Barnes, and Eusebio Scornavacca, "Consumers' Perceptions and Attitudes towards SMS Mobile Marketing in New Zealand," *Proceedings of the International Conference on Mobile Business*, 2005, 434–40.
11. Shayne Bowman and Chris Willis, *We Media: How Audiences Are Shaping the Future of News and Information* (The Media Center at the American Press Institute, 2003, http://www.hypergene.net/wemedia/weblog.php).
12. Xinhuanet, "3G Standard Test Will Be Finished between October and December 2006 in China," July 9, 2006, http://news3.xinhuanet.com/weekend/2006-07/09/content_4810569.htm.
13. Ministry of Information Industry of the PRC, "Quicken the Pace of Creativity and Transformation, Promote the Development of Telecommunication Better and More Quickly," March 22, 2007, http://www.mii.gov.cn/art/2007/03/22/art_2800_29524.html.
14. Jeremy Rifkin, *The Age of Access: The New Culture of Hypercapitalism, Where All of Life Is a Paid-for Experience* (New York: J.P. Tarcher/Putnam, 2000), 7.

16 Reinventing Newspapers in a Digital Era
The Mobile e-Paper

Wendy Van den Broeck,
Bram Lievens, and Jo Pierson

INTRODUCTION

This chapter focuses on how the newspaper as a medium is in the middle of a transition from an analogue "mobile" medium to a digital mobile medium. The newspaper has in fact always been a "mobile" medium, in the sense that readers could take this news medium with them on public transport, in the café, in the waiting room, or on other public places. In the first phase of digitization the newspaper got fixed to a desktop computer or at best to a laptop computer. But the next step is to "mobilize" the online newspaper, by giving the digital news content a "shell" that adequately enables everyday newspaper practices.

One of the more promising efforts is the e-paper device based on e-ink technology. This e-ink technology gives device manufacturers the opportunity to embed electronic screens with a very high resolution that looks like real paper, is ultrathin and, in the future, will also be flexible. This technology is amongst others being used by Sony, Philips, and iRex to develop mobile e-Paper devices (also referred as eReaders),[1] which are mobile devices to read newspapers, books, and documents with the same reading quality as regular paper.[2] The high screen resolution enables them to represent an interface that closely resembles the original newspaper artifact. In the trade-off between being digitally innovative by incorporating different functionalities and being recognizable as a (digital) newspaper, the e-Paper clearly tilts over to the latter. The e-Paper device is clearly designed to be as familiar as possible for analogue newspaper readers, with much less affordances[3] than integrated in current digital mobile media.

Our study investigates the acceptance and domestication of this kind of digital newspaper device by readers within their everyday life practices. The question is to what extent the familiarity with the classic newspaper really furthers the appropriation of this digital newspaper. Or do people expect that a digital mobile news medium should have as much functionality as possible, even beyond the standard newspaper reading?

In order to do this kind of research we first need a semimature technology that is, however, stable and fully functional in a real-life test environment. If the latter is not the case, then the first attention will go to these technological and interface issues. Only when this basic threshold is taken, a more contextualized and practice-oriented living-lab research can be undertaken.

Although the e-Paper devices we discuss within this chapter exist since a couple of years now, only few research projects with newspapers took place. So far only two commercial initiatives with mobile newspaper devices have been developed: *Les Echos* in France, and *The Yantai Daily* in Yantai, China. Both newspapers offer a daily edition of their newspaper on an adapted format for the iRex iLiad eReader.

In the light of these developments, we will discuss the possible benefits of mobile newspapers for users and the way they appropriate these new devices. The empirical data in this chapter are largely based on our own user research results within the Flemish IBBT project "e-Paper," in which we participated. Furthermore, within the Me-Paper project, we also examined other research projects, but due to their confidential nature, only few actual results could be gathered.[4]

E-PAPER NEWSPAPER RESEARCH PROJECTS

After an international search for R&D projects with newspapers adapted for mobile newspaper devices like e-Papers or UMPC (Ultra Mobile PC) in the Me-Paper project, we only found nine: *Les Echos*, *Yantai Daily*, *New York Times* Reader, *Herald Tribune*, LAT eMPrints, Diginews, Yomiuri, MINDS, and e-Paper. The research project "e-Paper" was the first trial in the world in which an electronic e-ink device was used to test a new mobile platform to distribute and consume newspapers in a real-life situation during a longer period of time. The objective of the project was to offer the possibility to newspaper readers to read their newspaper on digital paper in a user-friendly way (that is, e-Paper). In this chapter we focus on the user research that has been conducted, but also other issues were investigated (business modeling, advertising opportunities, technological research regarding, for example, architecture and terminal).

A sample of subscribers to *De Tijd*, the main Flemish financial newspaper, was invited to test the e-Paper, developed by Philips and iRex technologies. The total duration of the project was eighteen months, while the actual trial ran from April to June 2006. The research setting for the user study was based on the living-lab approach described following.[51] The specific aim of a living-lab user research is to provide developers with specific user insight on the e-Paper device, based on current user practices in a real-life setting. Next to that, by providing these devices to users in their natural environments, more fundamental questions could be investigated.

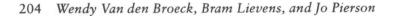

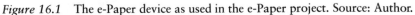

Figure 16.1 The e-Paper device as used in the e-Paper project. Source: Author.

WHY INVOLVE THE USER?

The e-paper project was set up within the approach of social shaping of technology (SST),[6] which means that the user was closely involved in the development process. Different technology failures in the past have shown that users and designers have a different way of looking at innovation. According to Frissen, the problem with innovation is that companies and developers all too often want to demonstrate their innovativeness.[7]

Users and technologies should be seen as coconstructed, which refers to the coconstruction or mutual shaping of social groups and technologies.[8] Technology is regarded as a seamless web of interacting, technical, social, economical, and political actors and factors.[9] However, for a long time the focus has been mainly on technological development and not on the usage of technology.

Several scholars in the STS field have made efforts to integrate the social constructivist view in technology usage.[10] An influential approach in this regard is the domestication school of thought, initially based on the work of Silverstone, Haddon, Hirsch, and Morley, and linked to cultural studies.[11] This approach signifies a fundamental break with technological determinism, but also with social determinism. It is the "mutual shaping" or mutual dependency between technology and social change that is taken into account, whereby technology influences everyday life and everyday use transforms the technology. Or, to put it differently, the domestication is anticipated in design, while at the same time the design is completed in domestication.[12]

The domestication perspective is not only about how users or potential users behave in relation to the technology and vice versa, but also about how people deal with ICT, which can also be an articulation of existing practices, conflicts, and meanings within the user community.[13] In order to understand the needs of users, recent studies on technology development argue that "Technologies should be studied in situ and in use, as part of socio-technical arrangements of humans and machines joined together in action and embedded in social contexts."[14]

As a result, the way people use a (new) technology is not always as expected or prescribed. As Nelly Oudshoorn and Trevor Pinch, identify "There is no essential use that can be deduced from the artefact itself."[15] Therefore, technologies should be studied in their own "context of use," and we should look into how users actually use a technology in everyday life. A second important aspect is the question what technologies do to users: "Working out who the new users are and how they will actually interact with a new technology is a problem familiar to many innovators of new technology."[16]

Another interesting aspect linked to this is "user innovativeness." Indeed, innovations reveal themselves in new social practices.[17] In some cases, the initiative to develop or use ICTs is initiated by users themselves, instead of by ICT companies or professional designers.[18] It is clear that involving users at an early stage of the development and design process can result in valuable insights in future user practices, and is a necessity in order to obtain a technology that fits users' needs and expectations.

This user innovativeness is important with regard to media and newspapers in particular. Research has showed that the evolutions in the media market have a direct impact on newspapers and the way those newspapers relate to their readers.[19] Newspapers and other media are being challenged by new emerging players as well as by the changing role of consumers. Due to converging technologies and new functionalities, consumers are becoming proactive. Not only the Internet but also interactive television enables people to become prosumers of news. Newspapers are directly challenged by these new evolutions. In 2006 *The Economist* titled "Who killed the newspaper?" questioning the viability of newspapers within the changing (mobile) media landscape.[20] The whole blogosphere, news aggregators providing RSS-feeds, have changed the user practices of news consumption.[21] The development of an e-Paper device, as an alternative platform for newspapers or e-books, can be seen as a consequence of this changing environment.

HOW TO INVOLVE THE USER?

Within the e-paper research project, a living-lab approach was applied, in which users were involved in the development process within their natural environment. A living lab can be defined is "an experimentation environment in which technology is given shape in real life contexts and in which [end] users are considered 'co-producers.'"[22] The living-lab setting can be applied in an early stage in the design and development process, enabling configurations on the technology level.

In our project 200 users received an e-Paper device they could use during two and a half months (April–June 2006). A living-lab configuration based on four consecutive stages was applied: contextualization, concretization, implementation, and feedback.[23] In the first two phases, the contextualization and concretization phases, the borders for the actual implementation of the test device are defined. The panel members were selected through purposeful sampling, an in-depth profile of each panel member is acquired, and then the device is introduced and handed over to the users. By analyzing the panel members' profiles and habits and usage patterns prior to the introduction of the new technology, the actual influence of the new technology on the specific media habits can be assessed.[24]

We conducted a series of focus group interviews (six focus groups with thirty-nine respondents in total) as well as three online questionnaires filled out by the entire panel. Based on these data, we were able to identify and validate what position this mobile device for reading newspapers could take within their everyday life practices. The final phase of the living-lab approach was the feedback phase in which the results were communicated to the development team. These findings were to be incorporated in the next versions of the iLiad eReader.

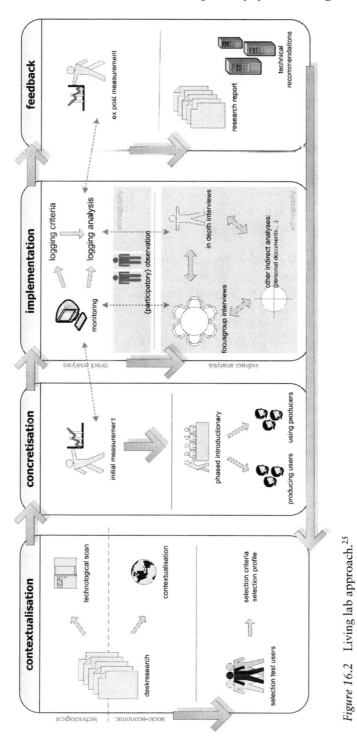

Figure 16.2 Living lab approach.[25]

THE MOBILE E-PAPER: THE NEWSPAPER OF THE FUTURE?

In order to answer the central question in this paper—the future position of a mobile e-Paper device based on users' everyday life practices—our main findings are presented in four parts. First, we identify the strengths and weakness of the e-Paper devices from a user perspective. Second, we investigate how a new type of distribution platform for content and news is positioned in the overall (news) content landscape. Third, we explore how the e-Paper device, which is seen as a mobile device, relates to other, more general mobile devices. Fourth, and maybe most importantly, we analyze the (new) emerging practices of the user.

STRENGTHS AND WEAKNESSES

The core technology of the e-Paper, which is the e-ink technology, produces an excellent screen quality (the e-ink screen) that was unanimously acknowledged by all test panel participants in the e-Paper project. The screen quality was rated extremely high since users considered the reading quality to be equal to that of "real" paper. In contrast to regular computer screens, the e-Paper causes little eyestrain when it is used for extended periods of time, and the screen is even readable in full sunlight. Because of its resemblance to real paper, the test panel perceived the e-Paper mainly as a reading device. In the focus groups, people imagined computers with this screen technology and foldable screens that were portable and on which people could take notes like on real paper: "If I could have this kind of screen on my computer, I would sign in immediately" (KS, male, thirty-four years old).[26] In a future evolution towards a paperless society, respondents see this as a possible useful alternative for real paper.[27]

As this project used a first demo version of the iLiad eReader, a number of technical issues limited the proper use of the device. The technological constraints at the start of the project—due to the early and less mature version of the e-Paper device—enabled us to identify the issues that are essential for having an optimal use experience. The instability of the device, slowness, short battery-life span, and problems with downloading the newspaper clearly had a negative impact on the use of the device as well as on the media consumption. Although some of these issues were tackled during the project, we noticed that no single respondent used the e-Paper device on a daily basis. This means that the device at that time was not mature enough to be an adequate media device. As one of the respondents in the focus groups (MDM, male, fifty-seven years old) argued: "If you want to read a newspaper, you want to know what's in it and you don't want to wait 20 minutes to see what's in it, then I have already read the regular newspaper. It was a waste of time to work with

Table 16.1　SWOT analysis.

Strengths	Weaknesses
Screen quality	Maturity of the device
Screen size	No clear added value
Compactness of device	Vulnerability
Look of device	Additional device (not AIO)
Perceived as reading device	Presentation of newspaper on the device
Long battery-life span	
Possibility to store a lot of content	

Opportunities	Threats
Multiple titles (newspapers, books, magazines)	Competition with other portable devices (PDA, smart phone)
Books on vacation	Habit of reading the daily newspaper (old habits die hard)
Reading device with Acrobat Reader or integrated devices (agenda, e-mail)	
Foldable screens	New/other actors
New ways of advertisement	Other mobile Internet-enabled devices
Personalization of content	
New business models	

the device." But despite these technical limitations, the research identified several opportunities and threats of the e-Paper as a new mobile platform for media—and newspaper in particular—consumption.

The other assets of the electronic newspaper on the e-Paper device and of the present design of e-Papers are summarized in the following Strengths, Weaknesses, Opportunities, and Threats (SWOT) analysis.

POSITIONING IN THE NEWSPAPER AND NEWS CONTENT LANDSCAPE

In terms of news consumption, people often use different sources. While the newspaper is mostly used for the "full story," radio is used to catch up with the latest headlines and television for the illustrative video material. Internet, as a full multimedia platform, is able to embed all those functions. Therefore, most newspapers also have Web sites where the newspaper can be read and the latest news facts can be followed during the day, even with additional audiovisual messages. Although many panel members connect to newspaper Web sites for updates during the day, on the e-Paper device, most people stick to the habits of their regular newspaper. The layout of the regular newspaper helps readers to make a choice in the articles they want to read: a long article can be postponed until later; only the first paragraph of some articles is read; small articles are regarded as less important than large articles, and so on. KS (male, thirty-four years old) explained:

> I choose the articles I read based on the titles and subtitles and the length of the article. Do I have the time and do I want to read a long article? And then I decide based on the entire page. In the newspaper you sometimes have articles of four lines and you have articles of an entire page. And that is important to know in advance, to know if you want to invest time in reading the article.

For people who are already used to consuming online news, an evolution towards an online newspaper on the e-Paper device with lists of titles and articles and a search function per topic is appealing, but for most readers the link with their regular newspaper is a necessity: "I want an exact copy of the newspaper in an electronic way, then I don't have to open my newspaper on the bus" (LVDM, male, fifty-eight years old). This is also linked to the fact that people subscribe to a specific newspaper because of the balanced content and its trustworthy character as a news source. The reader expects the same news aggregator function from the digital version.

Furthermore, readers not only expect the same functionality as in the printed version, but they also want to operate it in a similar way, at each level. Navigation, content, and structure should have the same level of comfort people are used to with their traditional paper. Therefore, the e-Paper has to enable some of the old habits associated with newspapers, like saving and archiving articles.

The interest in more Internet-related interactive functionalities such as personalized content, news alerts, or RSS feeds is, perhaps surprisingly, rather low. This is linked to the perception people have of the device as a reading device. They see it as just another way of distributing their familiar newspaper(s) and not (yet) as an online news-aggregating device. However, as e-Papers are further developed together with ubiquitous Internet access, we expect that the consumer will force the newspaper to offer the choice for a more interactive and personalized approach so that it aligns more with the Internet practices of news consumption (for instance, personalized news alerts, RSS feeds) of some of their readers.

POSITIONING AMONG EXISTING MOBILE DEVICES

Although the e-Paper is a portable device, it distinguishes itself clearly from other (multifunctional) mobile devices by an explicit focus on its reading functionality. The panel members could read newspapers, books, and their own documents (PDF and txt), view pictures and take notes. But what does the user expect from this device? How should the e-Paper be positioned between the mobile (media) devices that are already on the market? Does it need to be a complete mobile-media device?

We found different views on these issues. Businessmen in our panel, who already use mobile devices like PDAs and laptops, do not want an additional device. For them, the reading functionality should be incorporated in a PDA, for example, or the e-Paper should have the same functionalities as a PDA.

> You know those small tablet PCs in which you can manage your agenda and address book and mails? A small laptop, in fact. Then I would opt to put a newspaper on a tablet pc, if it's possible with the same screen quality as the e-paper. (WM, male, fifty-one years old)

> Today you want a device to have as many features as possible. I don't want to take my cell phone and another portable device and another portable device. (CV, male, thirty-six years old)

For other respondents, a mobile device with only a reading functionality is sufficient:

> From my point of view all those extra features aren't necessary. Those have nothing to do with reading a newspaper—those are computer utilities. (GV, male, fifty-nine years old)

> I prefer a device that is just really good [all]-in-one functionality. (PJ, male, twenty-eight years old)

This would replace the printouts on paper these respondents currently make. This group wants to read multiple papers on the same device and use the device to read their own documents in PDF. For this purpose, a tool such as Acrobat Reader needs to be incorporated.

READING THE E-PAPER: TOWARDS
NEW USAGE PATTERNS

Because of the prototype level of the devices it was not possible to identify specific news consumption behavior. However, in the in-depth focus group interviews, some new practices were prefigured. A first important remark is linked to reading books. On the e-Paper, several books could be stored, and some respondents also read books during the trial. The e-Paper is especially found to be practical to take books on holiday, as it is legible in the sun and it saves suitcase space and weight. However, the ability to download the daily newspaper on holiday was also found to be positive: "What was really nice was, right after I received the device I went on holiday to Italy. I just connected my e-Paper to the Internet and I received my newspaper. That was really nice" (DL, male, twenty-nine years old).

The aspect of habit is identified as a reason why people keep reading their regular newspaper. Reading a newspaper can be a ritual—especially on weekends, when more time is spent on newspapers.

It is not simply the innovation itself that is important to users. Rather, it is the (perceived) advantage of the technology they experience in everyday practices. Therefore, it is important to look at the specific benefits and added value of the e-Paper. We already mentioned that it is an excellent device to take on holiday or on trips abroad. However, in more everyday contexts the panel members mainly found it handy while on the move—in particular, for reading newspapers. This refers to public transportation, in which personal space is rather limited, and it is not always easy to spread out the newspaper. So the compactness of the e-Paper is a clear practical advantage, as it fits easily on the small train tables or even on the steering wheel of a car in a traffic jam.

CONCLUSION

As e-Papers and newspaper formats for the e-Paper are currently being commercially launched, the near future will point out what its specific impact will be on the everyday life practice of newspaper readers as well as for the global newspaper industry and the print industry in general.

In the short run, the e-Paper will certainly not replace the real newspaper, but for the future, there are some indications that e-Paper newspapers will become an additional distribution platform for publishing the existing newspaper. E-Paper devices offer the newspaper industry the opportunity to offer mobility and interactivity, without much adaptation of their present way of working. A lot will depend on the level of interactivity and personalization the consumer expects. Based on the e-Paper trial, we can expect that the e-Paper will not immediately lead to new reading patterns, as the user wants it to stay tuned to their reading patterns linked to the traditional newspaper. On the contrary, the trial showed that innovation through familiarity is a key element in the development of a new technology like the e-Paper. People expect the same functionalities as they are already used to with their existing newspapers. The connection with existing reading patterns of newspapers (and books) is therefore important and indicates that, at the moment, there are no new habits or behavior. Currently the added value is not on the functionality but on the portability and mobility of the e-Paper. The devices enable users to carry a large amount of books and newspapers on a very compact device.

In relation to other mobile media devices, the e-Paper causes no threat because of its explicit reading functionality. On the other hand, because the e-ink technology produces such a good screen quality, it could become an interesting functionality to build into existing mobile devices. So here an opposite effect could occur: the mobile devices could become a threat to the e-Paper in the long run.

Regarding the way that the e-Paper will position itself in the everyday life practices of the user, it is clear that, at least for now, the e-Paper will not have a major impact. In that regard, the e-Paper is (in first phase) not a new revolutionary device, but a substitute for existing platforms without a real influence on everyday life practices. But in the end, users will possibly transfer their existing newspaper reading practices to this new type of device. For newspaper agencies, the e-Paper enables them to bring their news on a mobile medium without any major adjustments on the organizational or editorial level. They can rather easily distribute the printed version on this e-Paper device, which potentially makes it very appealing for the newspaper industry. The e-ink technology has proven to be an outstanding technology for reading documents, books, and newspapers. It is a matter of time to see if a specific device with this technology, like the e-Paper, can become a full-grown mobile medium.

ACKNOWLEDGMENTS

This chapter is the result of research carried out as part of a research project funded by the Interdisciplinary Institute for Broad Band Technology (IBBT). The e-Paper project was carried out by a consortium of companies: Philips, iRex Technologies, De Tijd, Belgacom, Agency.com (formerly Hypervision), and i-Merge in cooperation with the IBBT research groups: IBCN, MICT (UGent), SMIT & ETRO (VUB), CUO & DistriNet (K.U. Leuven) & IMEC. SMIT (Vrije Universiteit Brussel) was involved in the user research in which it has applied its living-lab method. Project Web site: https://projects.ibbt.be/epaper. A second project this article relates to is the Me-Paper project, in which new formats for newspapers on electronic devices (e-Papers and UMPC) are being developed. Project Web site: https://projects.ibbt.be/mepaper.

NOTES

1. In this chapter we use 'e-paper devices' as a generic term to identify all types of devices that use e-ink as a core technology.
2. Examples of popular e-Paper devices at the moment are the Sony Libris, the Amazon Kindle, and the iRex iLiad eReader, although the latter is the only device that has been used for newspapers until now. Currently e-ink devices are still black and white, but color screens are predicted to appear on the market this year (Fujitsu FLEPia).

3. Affordances may be defined as the combination of "perceived and actual properties of the thing—primarily those fundamental properties that determine just how that thing could possibly be used" (Donald A. Norman, *The Psychology of Everyday Things* [New York: Basic Books, 1988] 95). See also: James J. Gibson, "The Theory of Affordances," in *Perceiving, Acting and Knowing: Towards an Ecological Psychology*, ed. Robert Shaw and John Bransford (London: John Wiley, 1977), 67–82; Wendy Van den Broeck, Jo Pierson, and Bram Lievens, "VOD: Towards New Viewing Practices?" *Observatorio*, 1(3) (2007), http://obs.obercom.pt/index.php/obs/issue/view/9.
4. See acknowledgment for more details on the E-Paper and Me-Paper research projects.
5. A separate research perspective concerning user issues was the usability research conducted by IBBT-CUO, a research center at KU Leuven. These results fall outside the scope of this paper.
6. Robin Williams and David Edge, "The Social Shaping of Technology," *Research Policy*, 25 (1996): 865–99.
7. Valerie Frissen, *De domesticatie van de digitale wereld* (Rotterdam: University of Rotterdam, 2004).
8. Nelly Oudshoorn and Trevor J. Pinch, eds., *How Users Matter: The Co-Construction of Users and Technologies* (Cambridge, MA: MIT Press, 2003).
9. Yves Punie, *Domesticatie van ICT: Adoptie, gebruik en betekenis van media in het dagelijkse leven: Continue beperking of discontinue bevrijding?* (Doctoraal proefschrift, Vrije Universiteit Brussel, Brussel, 2000).
10. Hughie Mackay and Gareth Gillespie, "Extending the Social Shaping of Technology Approach: Ideology and Appropriation," *Social Studies of Science*, 22 (1992): 685–716; Oudshoorn and Pinch, *How Users Matter*.
11. Roger Silverstone, Eric Hirsch, and David Morley, "The Moral Economy of the Household," *Consuming Technologies: Media and Information in Domestic Spaces*, ed. Roger Silverstone and Erich Hirsch (London: Routledge, 1992); Roger Silverstone and Leslie Haddon, "Design and Domestication of Information and Communication Technologies: Technical Change and Everyday Life," in *Communication by Design: The Politics of Information and Communication Technologies*, ed. Robin Mansell and Roger Silverstone (Oxford: Oxford University Press, 1996), 44–74; Leslie Haddon, "The Contribution of Domestication Research to In-Home Computing and Media Consumption," *The Information Society*, 22(4) (2006): 195–204.
12. Silverstone and Haddon, "Design and Domestication."
13. Jo Pierson, "Domestication at Work in Small Businesses, " in *Domestication of Media and Technology*, ed. Thomas Berker, Maren Hartmann, Yves Punie, and Katie Ward (Milton Keynes, UK: Open University Press, 2005), 205–26.
14. Matt Ratto, "Producing Users, Using Producers." Paper presented at *Participatory Design Conference*, New York, 28 November, 2000.
15. Oudshoorn and Pinch, *How Users Matter*, 2.
16. Oudshoorn and Pinch, *How Users Matter*, 2.
17. Ilkka Tuomi, *Networks of Innovation: Change and Meaning in the Age of the Internet* (Oxford: Oxford University Press, 2002), 251.
18. Haddon, "Contribution of domestication research."
19. Ike Picone, "Conceptualising Online News Use," proceedings of COST 298 conference, *The Good, the Bad and the Unexpected*, Moscow, May 2007, http://www.cost298.org/.
20. *The Economist*, "The Future of the Media—Who Killed the Newspaper?" international edition, vol. 380(8492) (2006), 9.

21. Picone, "Conceptualising Online News Use."
22. Pieter Ballon, Jo Pierson, and Simon Delaere, "Fostering Innovation in Networked Communications: Test and Experimentation Platforms for Broadband Systems," in *Designing for Networked Communications: Strategies and Development*, ed. Simon B. Heilesen and Sisse Siggaard Jensen (Roskilde, Denmark: Roskilde University, 2007), 137–66.
23. See also Bram Lievens, Wendy Van den Broeck, and Jo Pierson, "The Mobile Digital Newspaper: Embedding the News Consumer in Technology Development by Means of Living Lab Research," *Conference Proceedings of IAMCR 2006*, Cairo, 23–28 July, 2006; Jo Pierson and Bram Lievens, "Configuring Living Labs for a 'Thick' Understanding of Innovation," Conference proceedings of *EPIC 2005* (*Ethnographic Praxis in Industry Conference*), National Association for the Practice.
24. It is important to mention for the E-Paper case that the panel characteristics were based on the actual reading population of the newspaper *De Tijd*. As this is a financial newspaper, readership is male dominated (only 16 percent women), highly educated, and mostly situated in the upper social class.
25. Source: Jo Pierson and Bram Lievens, "Configuring Living Labs for a 'Thick' Understanding of Innovation."
26. For privacy reasons, respondents are referred to by their initials, gender, and age.
27. IBBT-SMIT E-Paper user report, 2006.

Part V
Mobile Imaginings

17 Face to Face
Avatars and Mobile Identities

Kathy Cleland

INTRODUCTION

With the growing pervasiveness of screen-based communication technologies, including personal computers and mobile phones, face-to-face communication is increasingly becoming augmented and in some cases even replaced by mediated screen-to-screen communication. With this growth of screen-based communication, the self is continually being mediated, remediated,[1] and intermediated[2] as it is networked and distributed through a variety of different media forms. We present ourselves as visual images on Web sites, in blogs, games, virtual worlds, IM windows and chat sites, and on mobile phones. As well as traditional indexical images such as photographs and videos, the avatar—a graphical 2D or 3D representation of the self—is increasingly becoming a familiar presence in online and computer-mediated environments such as games, virtual worlds, chat spaces, and now mobile phones. This chapter investigates how avatars are increasingly starting to act and interact as proxies for our physical selves, and looks at the sociotechnical forces shaping the design and use of these new avatar identities as they migrate from games and the Web to mobile phones.

Just as the Internet has become an increasingly image-rich environment since the introduction of the World Wide Web and improved bandwidth rates and download speeds, we are seeing a similar trajectory with mobile phones. Although mobile phones are still primarily used for voice communication and SMS, with increasing technology convergence they are now increasingly being used to access, create, and distribute a variety of image-rich media content. In countries like South Korea and Japan, where the data rates charged for mobile phone usage are relatively low and the transmission speeds are high, mobile phones are often used as the primary mode of Internet access. And, with new and improved convergent mobile devices like Apple's much-hyped new iPhone, released in mid-2007 in the United States, mobile phones look set to become even more significant devices for media production, distribution, and consumption, as well as for interpersonal communication including the use of avatars.

However, while the virtual body of the avatar offers some interesting and productive opportunities for new forms of technologically mediated identity and 'presentations of self',[3] it also represents an increasing consumerization and commodification of identity. Dominant sociocultural stereotypes (gender, race, class) are reinscribed on the virtual body of the avatar, both through the choices of users and via computer software programs where they become parameters and presets that perpetuate and reinforce existing social stereotypes and cultural norms.

AVATAR IDENTITIES

Virtual environment researchers Jeremy Bailenson and Jim Blascovich define an *avatar* as "a perceivable digital representation [in a virtual environment] whose behaviours are executed in real-time by a human being."[4] As the Internet has become increasingly media-rich, graphical environments and graphical avatars have largely replaced the earlier text-based environments and identities of role-playing games such as MUDs (Multi-User Domains) and MOOs (MUDs Object Oriented).[5] By assuming a graphical avatar, an individual can become digitally embodied in a virtual environment, and he or she can interact with other avatars and objects within that environment in real time. Research shows that using avatars as communicative proxies creates a strong sense of intersubjective presence and copresence in virtual environments.[6] This experience of real-time interaction has much in common with the feeling of 'shared presence' or 'presence at a distance' with which we are familiar in our everyday telephone interactions. With conventional telephony, this sense of shared presence is established by voice alone, but with the increasing media convergence enabled by digital technologies, mobile phones now combine the intimacy of voice communication with graphical signifiers of identity such as photographs, video, and avatars. On mobile phones it is becoming common to assign photographs as the caller IDs of friends and family or use them as screen savers and wallpapers. We also use our mobile phones to take and distribute photographs and videos of ourselves. When images are shared on a moment-by-moment basis, they can create a sense of shared experience that connects individuals in the visual reality of each other's lives even though they are physically separated. Mizuko Ito's investigation of mobile-phone use in Japan suggests that the sharing of photographs and viewpoints can create a shared "intimate visual co-presence" between groups of friends or intimate others such as boyfriends and girlfriends.[7] And because the mobile phone is carried with us everywhere we go, it enables individuals and intimate groups to maintain 24/7 this sense of "presence at a distance"—a "being here and being there," as Larissa Hjorth and Heewon Kim describe it.[8]

With the use of animated avatar images, real-time interaction and presence can be signaled graphically through facial expression, gaze, head orientation, and body movements, as well as by speech, which on the Web is typically represented in speech bubbles that appear above the avatar's head or in text boxes at the bottom of the image window. As avatars migrate to the mobile phone, however, text will be replaced by live speech, creating an even stronger sense of immediacy and copresence.

Although the avatar shares this quality of real-time interaction with other real-time images of the self, such as those offered by webcams, video-conferencing, and videophones, it offers two significant advantages. First, the animated graphical image of the avatar needs far less bandwidth than the real time video image, so that problems with image lag and jerky movements are decreased. This makes the avatar image a particularly attractive option for mobile phones. Second, unlike the video image, which is anchored in the physical reality of the body, the graphical avatar allows individuals to access more transformative modes of visual identity, enabling them to present a much more editable and customizable version of the self than is possible with the indexical video image. In this context it is interesting to note that one of the key reasons (in addition to technical constraints) that videophones have never really taken off is because they reveal too much of the individual's "backstage" environment. One of the key benefits of voice-only phone communication is that aspects of the individual's physical appearance and location can be successfully hidden. With the use of graphical avatars individuals can continue to successfully hide their aspects of their physical appearance and "backstage" environment while presenting a visual persona of their choice. Inappropriate clothing and/or locations and embarrassing behaviors such as blushing or sweating can be conveniently eliminated in the avatar presentation, giving the individual far greater control over the front stage "face" or "faces" they present to the world.[9]

Indeed, with the digital avatar, it is possible to be "whoever you want to be" (or at least look like them) without being limited by the physical specificities of your gender, age, race, or even species. In an online article "Get Real! Creating a Sim in *The Sims Online*™," Bob King, the lead artist for *The Sims Online*,[10] invites users to experiment with the look of their avatar identity by literally stepping into a new skin:[11]

> Have you ever wanted to know what it feels like to be someone else? Ever dreamed of the ability to step into a totally different skin than your own? Would you like the ability to create a persona for yourself that could have your social and communicative qualities, and yet look nothing like you? Imagine being able to walk up to a good friend, hold a complete conversation with him, and have him never recognize you. Would you like a skin-colour change? Have you ever fancied a gender change? Ever thought of becoming an alien? You can enjoy this type of charade and many more online in *The Sims Online*.[12]

Although these types of avatars are only just starting to emerge on mobile phones, they are already a significant presence on the Web and in games and virtual worlds; so it is instructive to look at the kind of avatar identities that are emerging in those contexts to gain some insights into the types of avatar identities that we may soon see more of on mobile phones and other portable convergent devices.

The promise of inhabiting a new skin and creating a fantasy alter ego or second self through the avatar identity lies at the heart of the popular Second Life® virtual world.[13] Rather than just watching our favorite media identities on television or in films, through avatars we can now become cartoonlike media identities in our own right as we interact with each other in the new media terrain of games and virtual worlds. While some Second Life® virtual-world residents, particularly those who have a recognizable and marketable real-life identity, inhabit avatars that resemble their offline selves (albeit frequently glossier and younger graphically rendered versions of themselves), one of the pleasures of online worlds is being able to construct an idealized fantasy identity and the lifestyle to go along with it. In Second Life® virtual world you can be a model or a rock star; you can wear designer clothes, drive in expensive cars, and live in palatial mansions.

Although maintaining a consistent avatar identity can be an important factor in establishing trust and building up a reputation in virtual communities, individuals often have a "wardrobe" of different avatar identities from which they choose to show different facets of their personality or to experiment with new identities.[14] For example, individuals may have a "work" avatar, an "intellectual" avatar, a "sporty" avatar, a "party" avatar, and so on.

AVATAR IDENTITIES AND PRESENTATIONS OF SELF

In *The Presentation of Self in Everyday Life* (1973), sociologist Erving Goffman describes the various "presentations of self" we enact in our everyday real-world interactions where we present different versions of ourselves in different social contexts.[15] Using a theatrical metaphor, Goffman describes these presentations as a type of performance, where we move between "frontstage" (public) and "backstage" (private) arenas, manipulating our appearance (clothes, accessories, makeup) and behavior to present different personas in different social roles and interactions—for example, student, daughter, girlfriend, friend, employee. This idea of individuals carefully constructing and performing their different social identities through a range of techniques of impression management provides a useful framework for analyzing the technologically mediated presentations of self that are currently being enacted on the Web and are starting to emerge in the mobile-phone environment.

In online environments, individuals can achieve much greater levels of control and "impression management" over their presentations of self than they can in their real-world interactions. In personal home pages,[16] blogs, and social networking sites like MySpace and FaceBook, photographs and text can carefully be selected and edited to present the individual's desired persona within that particular environment. With avatar identities, even more "impression management" is possible, as the individual can control and transform his or her visual identity beyond their real-life specificities of age, gender, race, and appearance. In online environments, the individual's real-world physical appearance and physical environment can remain completely hidden "backstage" so that only the desired "front-stage" avatar identity is visible. Individuals can also move seamlessly between different online "windows" or "social frames," where they can activate and play out different performative identities.

The difference between the "front-stage" online presentation of self and the "backstage" reality is the focus of a series of digital prints created by Australian-Malaysian artist Emil Goh (see Figure 17.1). Goh's *MyCy* (2006) series of digital prints is based on South Korea's hugely popular online community *Cyworld*,[17] which is accessed via mobile phone and the Internet.

Figure 17.1   Emil Goh, *MyCy*, digital print, 2005.

In *Cyworld*, community members create and customize their own cartoonlike avatars and mini-homepages. Each *Cyworld* member has their own miniroom, created by selecting various backdrops, wallpaper, furniture, and other appliances and accessories. The miniroom operates as the backdrop for the avatar and plays an important role in creating the avatar's identity. For many young South Koreans who still live at home and go out to socialize, the miniroom is their opportunity to create their own ideal fantasy living space which can express their tastes, aspirations, personality, and group affiliations.

In *MyCy*, Goh presents a series of prints which show the "front-stage" view of *Cyworld* avatars in their minirooms, paired with a "backstage" view showing the *Cyworld* members in their real-world bedrooms. The prints highlight the discrepancy between the fantasy self projected in the idealized avatar identity, with its aspirational lifestyle, and the individual's real-life identity and environment.

Goffman's analysis of front-stage and backstage behavior and the different social frameworks and environments within which social interaction occurs has also been widely applied in the study of mobile-phone usage.[18] However, in contrast to the clear distinction between front stage and backstage that we see on the Web, with the public performance of mobile-phone conversations this distinction is not so clear-cut. Mobile-phone users, particularly in public places like cafés or on public transport, typically have two audiences, the person on the other end of the phone and the audience that surrounds him or her in the physical environment. While the recipient of the mobile-phone call may only get to see the front-stage presentation, those in the immediate environment of the person making the call also get to see and share in the participant's backstage environment.[19]

AVATARS IN THE CROSS-MEDIA ENVIRONMENT

With the growing popularity of avatars, along with increasing media convergence and cross-media activity, we are starting to see avatars (or simplified versions of them) migrate from games and virtual worlds to other applications and platforms such as chat rooms, Internet messaging (IM), and mobile phones.

While a strict definition of *avatar* would limit its use to the description of animated graphical characters controlled by users in real time, in general usage the term *avatar* is also often used more inclusively to include a variety of still or minimally animated images that represent users in online environments such as home pages, chat spaces, IM, and mobile phones.[20] The popular Yahoo! avatars[21] used in their IM service are constructed by choosing different identikit facial features, wardrobe items, and accessories from a library of items. The avatars' facial expressions can also be controlled by sending emoticons so they can perform simple gestures and facial expressions.

Other examples of simple avatars that can be used either on the Web or on mobile phones are WeeWorld's WeeMee avatars[22] and Skype's Klonies avatars.[23] In the United Kingdom, the Glasgow-based DA Group and their consumer entertainment brand Yomego®[24] offer a range of animated avatars and virtual characters for television, the Web, and mobile phones. In 2006, Yomego launched MTV Flux, a new cross-media (TV, Web, mobile) virtual community where members can create an animated 3D Flux avatar that can appear on the MTV Flux television channel. Members can communicate with other "fluxers" by texting their avatars, and can request videos or upload media content on the MTV Flux channel either by SMS or through a Web interface.

However, while some of these avatar examples are being designed for use on multiple platforms and devices, in many cases competing proprietary standards and problems with interoperability will continue to make it difficult for users to have a "universal" avatar that they can port across different media applications or platforms. Your *Sims Online* avatar won't work in Second Life® virtual world or on your mobile phone and, although it is possible to export still avatar images and prerecorded animations and machinima[25] to different media platforms and applications, the real-time interaction of the avatar is only possible within the application in which it was created. Interoperability issues and the complexities of different mobile phone standards and platforms may also limit the uptake of avatars on mobile phones.

Creating viable mobile-phone avatars that can be controlled in real time depends on increased improvements in bandwidth and transmission speeds, as well as new software applications and hardware modifications. In the mobile-phone environment, there is also an increased emphasis on the spoken voice instead of the speech bubbles and text windows which are the norm in virtual communities and chat environments. New improvements in synchronizing avatar lip movements and facial expressions with real-time speech inputs (in addition to prerecorded voice and text inputs) are necessary for real-time avatars to be a success on mobile phones, and it is likely that these developments will in turn flow back into virtual worlds.

Companies such as Motorola Labs and California-based company 3-Dmsg are starting to develop more complex animated 3D avatars that can "talk" on behalf of users on mobile phones, and there is a lot of research going into developing applications that can generate real-time avatar-mediated messages by using speech recognition and video image recognition to generate the avatar's speech, facial expressions, and movement[26] (Jana 2006).

SeeStorm, a subsidiary of the Russian company SPIRIT, offers a range of ready-made and customized 3D avatars that can be animated by real-time voice inputs. The animation is generated by matching the avatar's mouth movements with the live voice input using speech phonemes recognition.

Figures 17.2 and 17.3 3D mobile phone avatars designed by Motorola Labs. Images courtesy Motorola, Inc.

Figure 17.4 SeeStorm's 3D talking head avatar for Skype phones. Image courtesy of www.seestorm.com.

Figure 17.5 OKI Electric Industry's FaceCommunicator®-BBE uses a camera and its FSE™ (Face Sensing Engine) to map users' facial expressions on to avatars.

The user can also change the avatar's facial expressions by selecting the appropriate smiley emoticon (see Figure 17.4).

In Japan, OKI Electric Industry has developed FaceCommunicator®,[27] a proprietary software application that enables users to control the movement and expressions of their avatar by using a camera to detect the movement of the users' eyes and eyebrows, and then using those inputs to generate synchronized movements in the animated avatar (see Figure 17.5).

COMMODIFICATION AND STEREOTYPES IN AVATAR IDENTITIES

While avatars offer some interesting and productive opportunities for individuals to control and craft their online identities, and to experiment with new identities, they can also be seen as representing an increasing consumerization and commodification of identity. In our media-driven consumer culture, where identity has become associated with physical appearance and possessions, consumption has become a primary site of identity formation as we seek to obtain the various identities that are promised along with the purchase of brand-name clothes and other consumer goods. In the virtual terrain, the selection of an avatar identity can be seen as just another consumer choice; in this case, however, as well as selecting clothes and accessories, we can also choose the more personal determinants of our identity such as gender, age, face, skin color, and body shape.

In Neal Stephenson's cyberpunk novel *Snow Crash* (1992), users who don't have the skill to design their own avatars typically buy ready-made off-the-shelf versions such as those described next.

Brandy and Clint are both popular, off-the-shelf models. When white-trash high school girls are going on a date in the Metaverse, they invariably run down to the computer-games section of the local Wal-Mart and buy a copy of Brandy. The user can select three breast sizes: improbable, impossible, and ludicrous. Brandy has a limited repertoire of facial expressions: cute and pouty; cute and sultry; perky and interested; smiling and receptive; cute and spacy. Her eyelashes are half an inch long, and the software is so cheap that they are rendered as solid ebony chips. When a Brandy flutters her eyelashes, you can almost feel the breeze.[28]

Versions of these Ken and Barbie clones along with other cute cartoon characters proliferate in games and virtual worlds and are also beginning to appear on the mobile platform. It is ironic (but perhaps not surprising) that, in this new virtual arena where theoretically we can be anything we want to be, homogenizing social stereotypes and idealized media types dominate the virtual landscape. These real-world norms and stereotypes are reinscribed in the virtual terrain both through the types of avatars made available for users by virtual-world developers and through the choices of avatars made by users themselves.

At the moment there are very limited options available for users to select or create their own animated avatars on mobile phones, but in the popular Second Life® virtual world, a thriving online marketplace has developed where users trade avatar skins, body shapes, clothes, and accessories. Users who don't have the skills to customize their own avatars can buy whatever attributes they desire, and in this marketplace sexy, idealized avatars dominate. Although there are a few offbeat "alien" or "monster" avatars, the most popular types of avatars are overwhelmingly made up of idealized Ken and Barbie clones and their updated popular culture cousins such as Japanese-style *manga* and *anime* characters. In Second Life® virtual world, it is easier to buy a glamorous idealized avatar identity than it is to design one that actually looks like your real-world self. These idealized identities are far more prevalent in the virtual terrain than in real life because they can be attained much more readily. Modifying your digital avatar is much easier, cheaper, and less painful than plastic surgery, and if you don't like your new look you can easily change it.[29]

The freedom of users to construct their own avatars is also limited by a number of design constraints. Designers and programmers make choices about the type of world or environment your avatar will inhabit, and they also determine the overall visual style of your avatar and its possible behaviors. Some of these decisions are a result of technical considerations (for example, hardware, software, and infrastructure limitations), but they are also the result of social, cultural, and aesthetic choices. As media theorist Vilem Flusser points out, our technological apparatuses are part of our culture, and "consequently this culture is recognizable in them."[30] Just as Flusser describes the photographer as the "functionary" of the camera because the actions are constrained by the functional abilities of the camera, so too are computer users the functionaries of the computer hardware and software systems they use:

> Photographers select combinations of categories . . . It looks here as if photographers could choose freely, as if their cameras were following intention. But the choice is limited to the categories of the camera, and the freedom of the photographer remains a programmed freedom. Whereas the apparatus functions as a function of the photographer's intention, this intention itself functions as a function of the camera's program. It goes without saying that photographers can discover new categories. But then they are straying beyond the act of photography into the metaprogram—of the photographic industry or of their own making—from which the cameras are programmed. To put it another way: in the act of photography the camera does the will of the photographer but the photographer has to will what the camera can do.[31]

Individual users or functionaries may challenge and subvert the cultural norms and stereotypes inscribed in the apparatus and its program, but only to the extent that the program itself allows; otherwise, as Flusser points out, they need to initiate change in the metaprogram itself. It is at this level of the metaprogram where the apparatus is constructed and programmed that sociocultural ideologies, practices, and norms are inscribed into the virtual terrain.[32] Different virtual environments have very different graphical styles or "looks," which range from cute cartoonlike avatars such as the WeeMees, Klonics, and Yahoo! avatars to the more highly rendered 3D avatars in Second Life® virtual world.

The users' freedom to personalize their avatar identity is a "programmed freedom" typically made by selecting from a predetermined array of mix-and-match options, body parts, and accessories. Users can generally change the gender and skin color of their avatars and select different facial features, clothing items, and accessories, but it is not always possible to change things like body shape or age. Slim, young (and usually white) adult is typically the default setting.

Even in more sophisticated virtual environments like the virtual world of Second Life®, where users are offered a vast range of options to customize their avatars' appearance, the slim, young adult is still the default setting. New users entering Second Life® virtual world are offered a choice between male and female versions of six default avatars. Until recently, the options included a Furry (a fox character), Girl/Boy Nextdoor, City Chic (young urban professional), Harajuku (an *anime*-styled character), Cybergoth, and Nightclub. Like the majority of avatar default types available online, all of these avatars (even the Furry) were young, slim, and white/Asian. The new default avatar selection has a few new avatar types; the Furry and Asian-styled Harajuku avatars have been removed and a black avatar

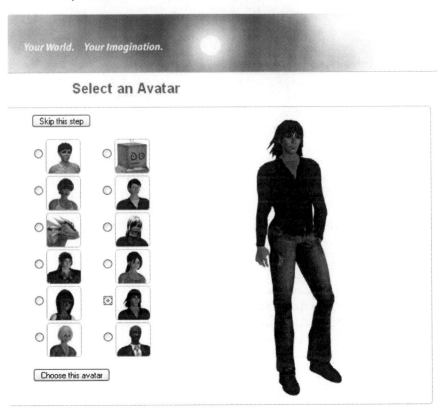

Figure 17.6   Default avatar selection in the virtual world Second Life®. Image reproduced with the permission of Linden Research, Inc. COPYRIGHT © 2001–2008 LINDEN RESEARCH, INC. ALL RIGHTS RESERVED.

has been added along with the nonhuman avatar options of a dinosaur and a cardboard box (see Figure 17.6). While these stereotypes may reflect the (presumed) preferences and demographics of users, they also serve to emphasize and privilege these particular identities and to suggest particular types of interaction and behaviors.

Your control over your avatar's movements, facial expressions, and gestures is also limited by the preset animations available within different environments—for example, waving, crying, blowing kisses, smiling, dancing, frowning, and so on.[33] These behaviors are typically quite stylized and exaggerated, and in some cases they are also heavily gendered—for example, if you want your Second Life® virtual world avatar to blow a kiss, you have to choose the male or female version of that action. In some cases, it is possible to create your own individualized movements and gestures, either by designing your own animations or by buying them from other players, but in many cases you are stuck with the preset options.

Although Second Life® virtual world residents are actively encouraged to modify the virtual environment and the appearance and behaviors of their avatars, this "freedom" is still constrained by the overall framework of the system and the design and programming skills of its residents. Residents who can't design and program their own avatar identities are limited to selecting from among the built-in features and items and behaviors available for purchase in the online marketplace.[34]

As demand grows for avatars on mobile phones it is likely that similar types of user-generated avatar designs will also start to emerge as long as they meet the technical specifications of different mobile platforms (and those specifications are made available to users and other third-party developers). But as we have seen with the Second Life® virtual world example, even when users can design or buy their own mobile-phone avatars rather than just choose from the limited selections offered by developers, it is likely that stereotypical avatar identities will continue to be dominant.

CONCLUSION

In the twenty-first-century media and communication environment, the avatar is set to become an increasingly common identity stand-in in our mediated screen encounters. In many ways, these avatar identities offer individuals far greater choice and control over their visual identities than they have with their real-world physical selves. However, as we have seen from the examples explored in this paper, embedded software preferences play a big role in setting the parameters for the appearance and behaviors of our new avatar identities, and also tend to intensify media stereotypes that idealize youth and beauty, creating an online world that is currently dominated by cute cartoonlike entities and hunks and babes. Even when users can go beyond the software limitations of avatar construction kits and design their own avatars and accessories, stereotypes still prevail, as is evident in the thriving online economies where avatar identities or skins, body shapes, clothes, and accessories are routinely traded and ideal types dominate. The "anything we want to be" would appear to be the stereotypical ideal types that proliferate in the popular media—young, buff, and good-looking with a full complement of materialistic accessories, designer clothes, hi-tech gadgets, cars, and palatial houses.

We can learn a lot about the likely types of avatar identities that we will see on mobile-phone platforms by looking at existing practices on the Web and in games and virtual worlds. However, as avatars increasingly start to migrate to mobile-phone environments, the norms and practices of the telephone medium will also have an impact on the style and channels of avatar communication. The importance of live voice will inevitably become a more dominant feature of avatar interaction than it has been on

the Web, where text (via speech bubbles and text chat boxes) still dominates. The combination of real-time avatar images with live voice interaction in future avatar developments for mobile platforms will produce an
increasingly stronger sense of virtual presence which blends the existing
strengths of the mobile and Web platforms.

NOTES

1. Jay David Bolter and Richard Grusin, *Remediation: Understanding New
   Media* (Cambridge, MA: MIT Press, 1999).
2. N. K. Hayles, *My Mother Was a Computer: Digital Subjects and Literary
   Texts* (Chicago: University of Chicago Press, 2005).
3. For a detailed discussion of Erving Goffman's description of "backstage"
   and "front stage" and the "face-work" individuals engage in to present an
   appropriate face in their social encounters, see Erving Goffman, *The Presentation of Self in Everyday Life* (New York: Overlook Press, 1973) and Erving
   Goffman, *Interaction Ritual: Essays on Face-to-Face Behaviour* (London:
   Allen Lane, 1972).
4. Jeremy Bailenson and Jim Blascovich, "Avatars," in *Encyclopedia of
   Human–Computer Interaction,* ed., W. S. Bainbridge (Great Barrington,
   MA: Berkshire Publishing, 2004), 64–68 (p. 64). In Hindu mythology, the
   term *avatar* is used to describe the material incarnation of gods when they
   take on a physical form to descend to earth and interact with humans.
   Just as Hindu gods manifested in many different avatar forms, so too can
   users take on multiple avatar identities in online and computer-mediated
   environments. The term *avatar* was first used to describe the graphical
   personas representing users in an online role-playing game called *Habitat* developed in the mid-1980s. See Chip Morningstar and Randall F.
   Farmer, "The Lessons of Lucasfilm's Habitat," in *Cyberspace: First Steps,*
   ed. Michael Benedikt (Cambridge, MA: MIT Press, 1991), 273–301. The
   term *avatar* then became widely popularized by Neal Stephenson's 1992
   cyberpunk novel *Snow Crash.* See Neal Stephenson *Snow Crash* (London:
   Penguin, 1992).
5. For more information about these text-based identities, which are important
   precursors to the graphical avatar, see Sherry Turkle, *Life on the Screen:
   Identity in the Age of the Internet* (New York: Simon & Schuster, 1995) and
   Julian Dibbell *My Tiny Life: Crime and Passion in a Virtual World* (New
   York: Harold Holt, 1998).
6. See Ralph Schroeder, ed., *The Social Life of Avatars: Presence and Interac-
   tion in Shared Virtual Environments* (London: Springer-Verlag, 2002); Jim
   Blascovich, "The Social Life of Avatars: Presence and Interaction in Shared
   Virtual Environments," in *The Social Life of Avatars: Presence and Inter-
   action in Shared Virtual Environments,* ed., Ralph Schroeder (London:
   Springer-Verlag, 2002), 127–45; T. L Taylor, "The Social Life of Avatars:
   Presence and Interaction in Shared Virtual Environments," in *The Social
   Life of Avatars: Presence and Interaction in Shared Virtual Environments,*
   ed., Ralph Schroeder (London: Springer-Verlag, 2002), 40–62.
7. Mizuko Ito, "Intimate Visual Co-Presence." Paper delivered at *UbiComp 2005:
   The Seventh International Conference on Ubiquitous Computing,* Takanawa
   Prince Hotel, Tokyo, Japan. (September 2005). Retrieved January 5, 2007,
   from http://www.itofisher.com/mito/publications/intimate_visual.html.

8. Larissa Hjorth, and Heewon Kim, "Being There and Being Here: Gendered Customising of Mobile 3G Practices through a Case Study in Seoul," *Convergence,* 11:2 (2005), 49–55.
9. Avatars can also provide an identity "mask" for users to hide behind. Not surprisingly, avatar identities are being promoted to marketing and call-center companies both to protect their operatives' identities and also to provide whatever face or faces might be deemed most advantageous for the company to present. See Reena Jana, "Building Your Own Phone Face," *BusinessWeek* online, retrieved January 19, 2006, from http://www.businessweek.com/innovate/content/jan2006/id20060119_883103.htm
10. See http://www.ea.com/official/thesims/thesimsonline.
11. Avatar identities are frequently referred to as "skins." These avatar skins, which include the gender, skin color, age, body shape, facial features, and hair color of the avatar, can be put on in and taken off in the online environment in the same way that we put on and take off clothes in the physical world.
12. Bob King, "Get real! Creating a Sim in The Sims Online™," retrieved October 1, 2003, from http://www.eagames.com/official/thesimsonline/features/fun_may02.jsp.
13. Second Life is a trademark of Linden Research, Inc. See http://www.secondlife.com.
14. See Sherry Turkle, *Life on the Screen: Identity in the Age of the Internet* (New York: Simon & Schuster, 1995); and John Suler, *The Psychology of Avatars and Graphical Space in Multimedia Chat Communities* (1996–2007). Retrieved February 5, 2007, from http://www.rider.edu/suler/psycyber.
15. See Erving Goffman, *The Presentation of Self in Everyday Life.* (New York: Overlook Press, 1973).
16. See Charles Cheung, "A Home on the Web: Presentations of Self on Personal Homepages," in *Web Studies: Rewiring Media Studies for the Digital Age,* ed. David Gauntlett (London: Arnold, 2000), 43–51.
17. Launched in South Korea in 1999, Cyworld (www.cyworld.com) currently has 30 percent of the country's population as registered users (eighteen million people) including over 90 percent of Koreans in their twenties. Cyworld has also launched sites in Japan, Taiwan, and China (two million users) and in July 2006 launched a test site in the United States. See Michael Kanelios, "Korean Social-Networking Site Hopes to Nab U.S. Fans," *CNET News.com,* August. Retrieved September 5, 2006, from http://news.com.com/Korean+social-networking+site+hopes+to+nab+U.S.+fans/2100–1025_3–6104794.html.
18. See Richard Seyler Ling, *The Mobile Connection: The Cell Phone's Impact on Society* (Amsterdam: Elsevier/Morgan Kaufmann, 2004) and Richard Ling and Per E. Pederson, eds., *Mobile Communications: Re-negotiation of the Social Sphere* (London: Springer, 2005).
19. See Leopoldina Fortunati, "Mobile Telephone and the Presentation of Self," in *Mobile Communications: Re-negotiation of the Social Sphere,* eds. Richard Ling and Per E. Pederson (London: Springer, 2005), 203–18; and Fernando Paragas, "Being Mobile with the Mobile: Cellular Telephony and Renegotiations of Public Transport as Public Sphere," in *Mobile Communications: Re-negotiation of the Social Sphere,* eds. Richard Ling and Per E. Pederson (London: Springer, 2005), 113–29.
20. Bailenson, Jeremy, Nick Yee, Dan Merget, and Ralph Schroeder, "The Effect of Behavioural Realism and Form Realism of Real-Time Avatar Faces on Verbal Disclosure, Nonverbal Disclosure, Emotion Recognition, and Copresence in Dyadic Interaction," *Presence,* 15(4) (2006): 359–72.
21. See the Yahoo! Avatars at http://messenger.yahoo.com/avatars.php.

22. See the WeeMee avatars at http:// www.weeworld.com.
23. See Skype's Klonies avatars at http://skype.klonies.com.
24. See http://www.yomego.com.
25. Machinima (machine cinema) are movies that are recorded in games and virtual worlds using in-game virtual cameras.
26. See Reena Jana, "Building Your Own Phone Face," *BusinessWeek* online, retrieved January 19, 2006, from http://www.businessweek.com/innovate/content/jan2006/id20060119_883103.htm.
27. See http://www.oki.com/jp/FSC/vc/en/bbe/index.html.
28. Neal Stephenson, *Snow Crash* (London: Penguin, 1992), 35.
29. Avatar skills and abilities are also commodified in the online marketplace. Many of the skill-based attributes of avatars in online role-playing games such as *Everquest* and *World of Warcraft* are developed within the game world itself and require a substantial investment of a player's time and game skills. A shortcut to acquiring an avatar with a high skill level and reputation is to buy an avatar with the desired skills or attributes from another player, and there is a growing market in avatar trading on online auction sites like eBay and PlayerAuctions.com, where avatars may be sold for anything from a few dollars to thousands of dollars. See Edward Castronova, *Synthetic Worlds: The Business and Culture of Online Games* (Chicago: University of Chicago Press, 2005) and Julian Dibbell, *Play Money: Or, How I Quit My Day Job and Made Millions Trading Virtual Loot* (New York: Basic Books, 2006).
30. Vilém Flusser, *Towards a Philosophy of Photography*. (London: Reaktion, 2000), 22.
31. Ibid., 35.
32. New media theorists such as Lev Manovich, Matthew Fuller, and Wendy Hui Kyong Chun write persuasively about how ideologies, stereotypes, and cultural practices are inscribed in computer hardware and software applications, where they perpetuate and reinforce existing social stereotypes, power structures, and cultural practices. See Lev Manovich, *The Language of New Media* (Cambridge, MA: MIT Press, 2001); Matthew Fuller, *Behind the Blip: Essays on the Culture of Software* (New York: Autonomedia, 2003); and Wendy Hui Kyong Chun, *Control and Freedom: Power and Paranoia in the Age of Fiber Optics* (Cambridge, MA: MIT Press, 2006).
33. These preset animations are activated by users clicking on an appropriate icon or, in more sophisticated environments, animations may be triggered automatically when certain words are typed or spoken.
34. In many ways, the nongraphical text-based MUDs and MOOs provided more freedom in terms of identity choice and behaviors, as users could be and do anything they could describe in words.

18 Re-imagining Urban Space
Mobility, Connectivity, and a Sense of Place

Dong-Hoo Lee

INTRODUCTION

Portable snapshot cameras have allowed people to visually record their experiences of specific moments and places. These technologies of memory, representation, and expression have shaped people's cultural practices of remembering and reproducing their life stories. When photographic images become digitalized and remediated via information communication technologies (ICTs) such as mobile phones and the Internet, and can easily travel in various communication settings of digital spaces, they constitute people's digital storytelling and their cultural significances are reinvented. The photographic images taken by portable digital cameras or cameraphones and transacted in the converged communication spaces of wired and wireless networks tend to transform what photographs have traditionally meant for people, how photography has been performed, and consequently how people perceive and make sense of the world.

As digital images created by users have proliferated on the Web, those that have captured people's spatial experiences have become one of main sources of creative online content. Especially when they have been linked to Web-based geographical maps, they have become an unprecedented source of geographical information. This study attempts to look at the ways in which urban experiences, captured by people's portable digital cameras or cameraphones, have been registered and constellated within the map on the Web. It examines a new form of geographical information created by ordinary people, something which has transformed the existing role of maps.

For this investigation, I will analyze *Cyworld* (http://cyworld.nate.com), one of South Korea's leading online social network sites (SNS), which is similar to *MySpace*. The site's map service has provided a platform for geospatial images created by the users. This specific instance will give us a venue in which to analyze the interconnection between people's photo-taking practices, their online activities, and the emerging geospatial imagery.

By analyzing the ways in which the expanding geographic information has been constructed, as well as how it has defined urban spaces and people's relations with them, I will discuss the role of technologies in affecting and even restructuring people's sense of place in the city.

THE SECONDARY "KODAK CULTURE"?: THEORETICAL CONCERNS

The so-called Kodak Culture or snapshot culture of ordinary people has been deeply intertwined with modern family life and tourism. The cultural processes through which the production and consumption of snapshots has been adopted as a constitutive part of the rituals of family and modern tourism reveal the cultural potentials of Kodak cameras as a modern mobile media, as well as their cultural conditions for actual usage.

To introduce easy-to-handle but poor-quality handcameras in the late 1890s and the early 1900s, Kodak emphasized the devices' playfulness and spontaneity in capturing outdoor experiences, and later, their ability to preserve domestic memories. In its campaign deployment, Kodak provided two contradictory images: the adventurous and fashionable "Kodak Girl," and the mother who compiles family histories via photographs. While the Kodak Girl symbolically emphasized the pure pleasure and adventure of taking photography via "a playful feminine spectator that ignores accepted standards of decency for the allure of freedom and mobility,"[1] it was soon substituted by motherly women who tried to be historians of family lives.[2] As the Kodak Girl's playful, individualist outdoor experiences paled in comparison to family leisure activities, snapshooting became a leisure activity which perpetuated and supported specific moments of family life, "chronicle[d] family ritual and constitute[d] a prime objective of those rituals."[3]

On the other hand, to carry cameras as "tourists' identity badges"[4] and to take photographs has become an indispensable part of the tourist experience. In this process, "travel becomes a strategy for accumulating photographs,"[5] since tourism advertisers and marketers, as well as mass media, publicize attractive images of the places. Commercial tourist imagery not only provides the pretext for tourism, which encourages tourists to be "looking at images or looking for images" rather than looking,[6] but also contributes to the so-called tourist's gaze on the destination.[7] Although tourist photography cannot be fully reduced to the preformed or framed performance,[8] snapshooting has constituted the foremost activity for the traveling public, and has shaped its spectatorship in relation to the places visited. While the existent familial ideology has been embodied in the popularization of snapshooting in the private sphere, the capitalist commercial projects have shaped the activities of snapshooting in public spaces.

Digital cameras and cameraphones seem to accomplish and intensify the early promise of the Kodak camera that emphasized the pleasure of the adventurous people on the move. As these devices are incorporated into everyday lives and are likely to be carried in people's pockets all the time, these personal and ubiquitous media extend the playfulness of photo taking and naturalize the "picture-thinking" of the Kodaker who "has his eye out for 'likely' subjects wherever he happens to be."[9] People can record the moments of their everyday lives and scenes they witness on the move, making the world in private and public spaces more visible and transparent. The intimate presence of cameras can easily allow photo-taking and photo-preserving activities to move beyond the ritualistic photographic practices bounded by traditional family narratives.

While the digitalization of photography has transformed the ways in which people take, print, and store photographs in albums, it also enables them to enjoy being more active producers and distributors of photographic images. It even intensifies the playfulness of snapshot photography; because of dematerialization, people do not have to worry about the costs of photo taking and photo printing. They can instantly check what they have photographed and easily delete unwanted images; thus, they are able to get sheer pleasure from photo taking itself. Several studies on the uses of digital cameras and camera phones show the degree to which photo taking has become a playful, everyday event rather than a meaningful ritual designed to record and commemorate the past.[10] With digital cameras and cameraphones as personal media of mobility and portability, people can take photographs that are less tied to their conventional social functions. They capture moments and places that conventional photography has typically disregarded; moments which are trivial and personally intriguing, and routine places where people do not usually carry cameras.

However, the changed materiality of photography has transformed the nature of photographic performance. The instant production ("play") and reproduction ("copy") of photographs and their connection to the wired or wireless Internet have affected the ways in which photographs are communicated. Photographs can be used to increase people's mutual experience in the photo-taken moment, and to help people remember, as well as to discuss, an experience later.[11] The reasons for photo taking and sharing include not only recording and recalling individual and collective experiences, but also creating and maintaining social relationships, and to express or present oneself.[12] While conventional snapshots have played the role of visual aide-mémoire through which people share images and tell stories, digital photographs provide more than memory and evidence. Digital snapshots come to have different cultural meanings because of the changing communicational contexts in which photographs are taken and shared. The oral condition, especially, which presides over the conventional processes of remembering and interpreting snapshots, has been transformed due to ICTs.

As conventional snapshots have typically been taken and compiled in a "home mode of pictorial communication,"[13] and shared in oral communication, they are likely to be shown to a limited circle of intimates, to function as a vehicle to narrate family history and generational lineage, and to confirm social relationships and communal membership. However, digital photographs can be shared both immediately and later—both in face-to-face and telecommunication settings, and by intimates, remote people, and even the unknown public. While people used to go through a "time-consuming and complex filtering and arrangement" of conventional photographs to fit them for their oral performance, they pay little attention to the archiving and compilation of digital photographs.[14] The oral performance, which used to be a post-photo-taking event to revocate memories, now proceeds in different ways. One can take photos and use them for ongoing conversations with people sitting together or located remotely; in turn, this oral performative condition can constitute a playful context for photo taking.

Moreover, as photo-sharing practices can be placed in more diverse communication modes, snapshots and their stories are extended beyond private or familial communication circles and can take on various social functions and meanings. The photographic album as "an instrument of collective show and tell"[15] and a generator of conversation is remediated by personal handsets and the wired or wireless Internet. As a result, its oral framework is flexibly constructed and reconstructed in various communicational contexts. The changing media environments affect how people talk about and share their photographs, and further, these photographic practices shape people's experiences and perceptions of the world.

SHARING PICTURES AND MAP IMAGERY

Photographs taken by individuals have become an integral part of user-created content on the Internet. As they are displayed and circulated via blogs, personal Web sites, or community bulletin boards, these personal expressions can easily be accessed by the public. Private photo-taking practices come to have new social and public meanings, and are potentially affected by the expansion of communicational contexts in which they are appreciated and interpreted. The dialogues and social relationships on the Internet can condition photographers' expectations for their outputs, their relationships with objects to be photographed, and thus, their experiences in the physical places. The maintenance of the photo-blog tends to provide motivation for taking photos.[16] Moreover, as Web activities like blogging become a major way of sharing pictures, they come to constitute another kind of oral performance that makes sense of and signifies photographs. Thus, Web activities tend to be implicated in the photographic act of framing, taking, displaying, compiling, and reviewing photographs.

Recently, images in photoblogs and photo-sharing sites that are linked to map services on the Web have become another resource for geospatial information. For instance, at geoblogger.com, one can see numerous push-pin-like place markers on a Google Map, which are linked to the popular photo-sharing site Flickr.[17] Images which are "geo-tagged" with latitude and longitude information come to provide a virtual sense of the designated place on the map.

The term *map* literally refers to "a diagrammatic representation of an area of land or sea showing physical features, cities, roads, etc."[18] While the map has been a cultural medium that represents people's imaginative geography for thousands of years, the "scientific map" that has tried to objectively represent the reality of territorial spaces and fix it on the atlas emerged only in the modern age. It became a part of everyday life and the national practices when early modern print culture and nationalism emerged.[19] Although many studies argue that maps constitute a socially constructed "rhetoric" or imagery contingent to historical contexts and complex sets of social relationships,[20] the popular belief in the map as a mirror of geographical reality remains firm.

Recent map productions and reproductions have been drastically affected by the development of the GIS (Geographic Information System), which digitally analyzes and processes geographic data. The computerized imaging and mapping practices of this system have challenged "the ontotheologies of realism and representationalism" of modern cartography, which refers to a belief in maps that exactly reflect nature and, thus, are fixed and unitary.[21] Especially when various geospatial sources are integrated into maps on the Internet, the map can be "an important part of search engine," and an index or "an interface to other geographic and nongeographic information on the web."[22] As search engines and indexes, maps can be a site where geospatial information is conglomerated and presented in various dynamic forms, allowing greater user interactions that are not possible with the unitary modern map. Such information and interactions can provide a chance to expose spatial identities or geospatial information that modern maps have excluded or obscured.

Many scholars have asserted that people's geographical imagination has been significantly shaped by the ways in which photographs have represented places and landscapes within their frames.[23] These geographical imaginations have usually been shaped by commercial photography, mass media, and occasional tourists' photography, which have produced and reproduced a series of preconceived notions about places. However, individuals' photographs, which have become one of the visual geospatial sources integrated onto maps, have begun to constitute recent "imaginative geographies." As they have become a source from which more and more people get their sense of place, these photographs tend to twist the photograph's role for "picturing places" and its contribution to the construction of imaginative geographies. Consequently, it is worthy to ask whether they merely reproduce the existing imaginative geographies and their inscribed relationships with the places, or transform the ways in which people make sense of the world and engage with them.

IMAGINED TOPOGRAPHY: *THE SOUTH KOREAN CASE*

In 2004, 24.9 percent of South Korea's population of 48 million people subscribed to broadband Internet services, which ranked at the top among the thirty Organisation for Economic Co-operation and Developmen (OECD) countries.[24] As of June 2007, 75.5 percent of the South Korean population aged six and over used the Internet, while 77.8 percent of its population carried one or more mobile handsets.[25] Moreover, 89 percent of handsets sold in 2005 were equipped with multimedia functions, including digital imaging. One of the noticeable cultural phenomena in South Korea since 2000, related to the ICTs, has been the development of personal social networks via personal weblogs or mini-homepages, as well as user-created content (UCC) sites.

According to the 2007 survey, 40.0 percent among the South Korean Internet users managed their own blogs or mini-homepages,[26] and especially, the users in their twenties actively lead this cultural trend. For instance, Cyworld's mini-homepage service, which started in late 1999, was popularly received among this population, and its usage rate had already reached 92 percent by 2004.[27] Cyworld provides blogs to allow people "to document their everyday life for friends and family" and "to forge familiarity and connection with friends but also operate as vehicles for meeting new and like-minded people."[28] It is a site where individuals set up profiles, post journals, and display creative works, in order to maintain and create personal relationships and to express and present themselves. Since the contents of these mini-homepages have accommodated photos as their substantial element, their popular reception has created or been accompanied by youngsters' enthusiasm for digital cameras or camera phones. A South Korean manufacturer's development of megapixel camera phones,[29] as well as the recent growth of digital single lens reflex (DSLR) usage, especially among young women,[30] can be construed as an offshoot of people's intensive use of photographs and their demands for higher-quality photographs to enhance their communicative ability on the Web.

The creative content produced by bloggers or mini-homepagers has increasingly been recirculated in more diverse, interconnected ways. Cyworld's map service (www.map.cyworld.nate.com), which opened in August 2006, has provided one of the sharing sites that recontextualize individuals' blog-based contents under specific themes or categories, in this case, to render them as another form of geographical data. This introduces an additional method of circulation or consumption path through which one's personal expressions can be displayed. When the users post personal experiences and expressions associated with various locations and places, they can attach the map's geographical information and post them to the so-called Story Map. The map provides people's daily updated images and "stories," and thus, one can obtain not only geographic information of a specific location on the map, but also representations of others' actual experiences there.

The map does not merely display pregiven geographical information, which usually reflects administrative and commercial interests, but also gives concrete, lively information produced by ordinary people. These users' contributions to the map service tend to affect people's expectations and senses of the map and consequently the concept of the map itself. The following section analyzes how this map represents place, paying attention to the role of photographs in representation. For this analysis, it sketches not only the characteristics of geographical representations but also the discursive structure of the Web page and in, particular, focuses on the people's story map at Cheonggyechon, the newly uncovered stream which runs through downtown Seoul in South Korea.

FRAMES VERSUS PERFORMANCES

The front page of the Cyworld map service consists of several nodes which are connected to geographical search engines and various thematic collections of geographical information. On the left, there is a shortcut to the people's story map and the newly listed stories, while on the right, there is a shortcut to the map search engine, where one can get road directions and traffic information, selected photogenic shots, a list of new popular places, and a list of regional members who have actively posted their stories. In the middle, one can see selected photographic images and location information categorized by specific topics or themes (see Figure 18.1).

The tourist-magazine-like front page includes various functions. While the service provider encourages users' participation and continues to attract them by selectively highlighting specific content created by users, users can get the pleasure of viewing the assorted geographical images and information, navigating the map, and peeking at what others have experienced in specific places. A user can find his or her own

Figure 18.1   Front page of Cyworld's map service (February 4, 2007).

work listed on the Web and, moreover, can use this service to improve the reputation of her or his own mini-homepage. When one looks at a specific area on the map, one can see both the popular list of stories marked in that area and a thumbnail image, popping up on the map, which is super-linked to the author's mini-homepage (see Figure 18.2). While the map itself consists of road grids and icons of public buildings or locations, the Story Map works as an archiving site, where individually taken photographs and personal stories are collected and shared. At the same time, it acts as a gateway directed to personal picture albums and webzines. The links created are bidirectional, not only connecting private content to a public map but also linking a public map to a personal social space where authors and readers can interact in terms of comments and feedback. The map functions as an assembly of personally created content, which in turn transforms the fixed, abstract notion of a map into that of a living, collectively constructed, and picture-oriented information conglomerate. Individuals' photo-taking performances on location and their posting activities turn ordinary personal experiences into shared public resources, and consequently widen the map's meaning and function.

Figure 18.2   An example of the Story Map.

UNEVEN SURFACE

Popular places, where photographs have been piled up on the Story Map, are not much different from the tourists' sightseeing attractions. For example, in the Seoul metropolitan area, those places which attract the most attention are most often photographed; the photos posted include: parks such as the Sky Park, Seonyudo, and the Children's Grand Park; urban landmarks such as the New Seoul Tower, the Cheonggye Plaza, and the World Cup Stadium; and historical or traditional sites such as the National Museum of Korea, the Gyeongbok palace of the Chosun Dynasty (1392–1910), and the Insadong district. These places are not merely the most frequently visited tourist destinations in Seoul but also the most photogenic locations in that they please people's visual appetite.

While two to five hundred pictures have been posted of these well-known tourist places, the moderately popular places, where twenty to two hundred pictures have accumulated, include: privatized public places such as shopping centers, restaurants, hotels, and theaters; public institutions such as universities, high schools, and subway stations; and those streets most populated with youngsters. There are also representations of moderately popular historical and contemporary landmarks, parks, and public places.[31] Private shops and restaurants, which are not usually registered on a given map, appear as places of interest. A large number of pictures accumulated for a specific site speaks to its popularity or reputation, and enlists it as a must-visit place. Moreover, those places where people meet and dine together, study and spend their school days, or pass by during their commute become noted as attractions where people are willing to take pictures; this marks them as notable places and draws public attention to them.

People still mostly tend to get their photographic objects from those which have drawn popular attention in terms of their historical and cultural significance, natural beauty, or consumption infrastructure. People's geographical imaginations have been unevenly accumulated in parallel to the social reputations of places and their photogenic qualities. Some districts in Seoul where no single place is represented by more than twenty pictures are either residential areas or factory sites, which have few if any famous landmarks, spectacular landscapes, major public or cultural institutions, or highly developed consumption spaces.

However, the Story Map conveys a wider range of geographical images and information than any previous map could include or denote. Through this map, one can view not only various and detailed images of a specific site but also be exposed to images of unexpected places which have almost not been considered to be photographable objects. Even those districts, which have few popular sightseeing places that would evoke people's photo-taking activities, are not totally neglected by the members. Images of a local park, playground, church, school, office, restaurant, apartment, subway station near the end of the line, supermarket, and so on, constitute the "long tail" of compiled photographs of Seoul. There is certainly a distinction between centers and peripheries in people's photographic attention; yet places usually ignored by mapmakers, tourist guide texts, or the popular geographical imagination become marked through individuals' subjective photo-taking practices. Every place and moment is waiting to be photographed or visualized, and the delineations of geographical reality are becoming more multilayered. The following section will further pay attention to the spatial portrayals of Story Map's photographic geographical imagination, especially focusing on images on the Cheonggye Plaza, the meeting point of Cheonggyechon.[32]

COLLECTIVE COLLAGE

Cheonggyechon is a historical site that symbolically demonstrates what Seoul has gone through during its compressed modernization. At the end of 1950s, the city government started to cover it up with cement under the slogan of the nation's urban development, and in the 1960s and 70s built an expressway and modern shopping districts over the covered Cheong-gyechon district. During the 1970s and 80s, this district was a leading location for Seoul's industrialization and yet, later in the 1990s, it became a stagnating downtown slum that could not keep pace with the rapid transformation of the cityscape under the increasing influence of globalization. When Myung-Bak Lee,[33] a former entrepreneur, was elected mayor of Seoul in 2002, his first ambitious project was to "clean up" Cheonggyechon by eliminating cement and expressways over Cheonggyechon and restoring the once buried stream. Although there was a great controversy about his project to recover a 5.8-kilometer stretch of the stream,[34] the project was completed on October 1, 2005.

Cheonggyechon, renovated by highly designed landscape architecture, has become one of the city's best attractions. Its changed outlook has not only transformed the cityscape at the heart of Seoul but also people's perceptions or experiences of it. As an inner-city park at the center of the metropolis, it functions as a resting place for commuters working in nearby companies and public offices, as well as citizens residing in metropolitan regions, a tourist spot where a sightseeing program is in operation, and a popular spot where amateur photographers take pictures. While domestic and foreign tourists visit this space to view a widely recognized urban landmark, city dwellers can easily access it during their lunch or break time. The splendors of newly built river walks, bridges, and other structures have drawn the attention of various groups of people, and have been captured by their photographic practices. The Story Map, where people upload what they have photographed in the Cheonggyechon area, provides us with a chance to witness the altered relationship between people and space within this area, and to look at what people experience with and in it.

As most of the spectacular structures are concentrated at the Cheong-gye Plaza area, this area is the most photographed location in comparison to the other parts of the Cheonggyechon. Waterfalls, river walks, waterways, and artistic structures, as well as seasonal festivals such as Luce Vista, the festival of lights, and the annual festival celebrating its restoration, have attracted photographic attention. More than two out of three pictures show night scenes at the Cheonggye Plaza. Phantasmagoric scenes, created by colorful lighting that illuminates the waterways, waterfalls, and other structures, provide people with attractive photographic moments. In addition, Luce Vista, held annually since 2004 during Christmas and the New Year, provides another spectacular illumination that inspires people to take out their cameras, visually capture its oneiric moments, and appropriate these scenes for their self-expressions on the Web.

By embroidering urban night scenes with its colorful brightness, and thus creating another dimension of cityscape, electronic light has intensified the voyeuristic experiences of urban streets. Spectacular lighting in the Cheonggye Plaza displaces the Cheonggyechon's historicity as a grand urban drain or a mark of urban industrialization with a trendy and exotic surface, and displays the type of urban spectacles that can be found in other global cities. People's photographs confirm that this area's main attraction lies in its ahistorical beautification and fantastic visuals, as well as its urbanness and globalness.

While the landscape around the Cheonggye Plaza is a main photographic object, people are also often photographed. Since the area is always populated by tourists and city dwellers, it is not easy to frame pictures without including people. Even some pictures straightforwardly foreground river walkers as a spectacle at the Cheonggye Plaza. Not only structures, but also crowds, become objects for people's visual pleasure and photo taking. As photographic practices constitute major activities within the area, it becomes natural to experience one's surroundings through a photographic gaze. Yet this photographic gaze is not homogenous. Some explore the camera's aesthetic effects to create stunning images, and others capture playful or unusual moments that many would miss. Private photographs of family members, friends, and loved ones, as well as self-portraits, are often posted. While aesthetic images of the place show its visual impressions, private portrayals of personal relations and private memories also slip into the photographic archive of the public Story Map (see Figure 18.3).

Even the attractions that draw people's photographic attention most frequently are not photographed in the same way. While pictures may show the same subject matter, there is no identical perspective. Camera angles, exposures, distance from the object, lenses, lighting conditions, photographers' positions, and so on, create unique images. Each photograph, from panoramic views to fragmented close-ups that try to capture the entities of photographic objects, reveals an individual's aesthetic sense, technological ability, and experiential relationship with the place at the time the photo is taken. Captions accompanying the photograph further explain the personal purposes of photo taking, photographic conditions, spatial experiences, and the degree of emotional involvement of the photographer. Each photograph, which more or less conveys subjective experience and has an exclusive quality, is put together to form a collective collage about a specific place. From the emulations of artistic photography to the traditional practices of family photography, from playful searches for photographable objects to emotional expressions of being there, and from rediscoveries of everyday urban space to tourist exclamations of scenic beauty, people's bits of experiences, which reflect their experiential interactions with the place, create the story-making narrative of the Story Map and characterize its geographical imagination.

Figure 18.3   Examples of photographs posted on the Story
Map. Source: Author.

SPATIAL MEMORIES IN PROCESS

On October 1, 2006, "Spring," the 21-meter artistic structure created by
Claes Oldenberg, was built in the Cheonggye Plaza to celebrate the first
anniversary of the plaza's opening. Since the city government's announce-
ment of its choice of the artist in early 2006, there has been harsh criticism
from local artists and civic activists due to the city government's closed and
unilateral decision-making-process, the artist's lack of understanding of the
historical specificity of the site, and the structure's symbolism in its lack of
association with local culture.[35] However, ordinary visitors do not seem to
care about the controversy; rather, they welcome another visual attraction
to be photographed, taking pictures of its grandeur and monumentality.

Photographic practices, in pursuit of visual spectacle or personal memory-making, tend not to reflect the theoretical discourse or historical background around the place or cultural rumination and statement about it.
Yet when individual photographs are gathered together to form a collective
collage of the place, they display not merely subjective expressions illus-
trating visual attraction but also collective visual memories which are still
in progress. Although individual photographs generally do not convey the
temporal dimension previous to the restoration of the Cheonggyechon or
the contextual dimension about the city government's controversial policy,

they provide records of what people are doing in and with the site, and how they remember its festivals, events, performances, and seasons. Spectacles of illuminated waterfalls, structures, Luce Vista, the river walks' seasonal changes—even during flood season, various performances and festivals, and people who are walking, taking pictures, playing, waiting, and engaged in activities with loved ones, as captured by cameras, give us a certain temporal dimension, as well as telling us about what relationships people have and how they try to remember.

Photographs posted on the Story Map are not chronologically arranged. People upload not only pictures taken "now" or recently but also what they find in old photo albums or files. Although the Story Map was launched in mid-2006, it is not difficult to find photographs on the opening of the recovered Cheonggyechon and other events before this time. The Story Map has absorbed every kind of record and memory that is associated with the place and has displayed this archive in a random array. The place's fragmented and polysemic visual records, which convey individuals' micronarratives, have been gathered together to form its social memory.

CONCLUDING REMARKS: A NEW SENSE OF PLACE

As cameras have become everyday personal belongings, and photo-taking practices are connected to various communicative activities on the Web, photographic images taken by individuals are increasingly proliferating and provide unprecedented sources of created content in digital space. The technological environment, which intensifies the playfulness of portable cameras and remediates oral performances related to photo albums, reshapes people's photographic practices, that is, the ways in which people take and make meanings out of photographs. In particular, as individuals' photographs become associated with Web-based geographical maps, they are profoundly engaging in how we perceive, orient ourselves in, and talk about physical space. They contribute to forming a new sense of place and, further, to making us perceive and interact with space via photographic eyes. Photographs shared on the Web not only provide visual images of places but also reveal the degree to which communication technologies have affected practices of "picturing places" and their applications.

The case of the Story Map, an archive of geo-tagged photographs, exemplifies how personal experiences and expressions about places have become a constitutive part of geospatial information on the Web. This case shows that geographical imaginations constructed by people's photographs are not fixed or bounded—but are rather always in process. Rather than creating a single place-myth, they reflect people's experiences and expressive abilities, and provide a loosely collaged image of a place, which collectively materializes individuals' perceptions and engagements with the place. Moreover, as photographs, used to elicit personal narratives of an individual's Web page or blog, are recycled to constitute geospatial imagery for the public, their geographical imaginations often include personally nuanced and emotionally laden expressions which were mainly absent from the modern map's diagrammatic representation. People's aesthetic explorations of or affective relationship with a place, and their personal commemorations of being there, continuously renew photographs' spatial indexes to conjecture not only the physicality of but also the spirit of the place, and reterritorialize the given map's geographical imagination.

Although the geographical imagery constructed by individuals' photographs usually follows visually attractive scenes or spectacles, and thus is likely to be susceptible to commercial or administrative efforts to colonize the cityscape for those ends, it is not totally subsumed by manipulative discourses to celebrate commodified space and to lead specific spatial symbolism and identity. While people are drawn to spectacular appearances and illuminations, their photographs do not merely collect the predetermined imagery but highlight individuals' experiential, playful, and aesthetic interpretations of a place. Each photograph's unique visual "story" consists of a multifaceted and ever-changing image of a place. People's visual memories of the present and the past of the place are interwoven onto the map, which turns into a multi-layered geographical image archive. As mobile media and their interconnection with the Web extend the individual's capacity for self-expression and social communicability, ordinary people can contribute to the construction of spatial imagery and its interpretations. The mobile public's media practices, including photographic practices, ask us to further explore how human subjectivity and the cultural identities of places have been rearticulated by the mobility and connectivity of today's media environments.

NOTES

1. Eric Gordon, "City in Play: Kodak, Cinema, and 'the Great White Way,' " presented at the *2006 Urban Communication Foundation Pre-NCA Seminar*, San Antonio, TX, November 15, 2006.
2. Nancy West, *Kodak and the Lens of Nostalgia* (Charlottesville: University of Virginia Press, 2000).
3. Marianne Hirsch, *Family Frames: Photography, Narrative, and Postmemory* (Cambridge, MA: Harvard University Press, 1997), 7.
4. Richard Chalfen, *Snapshot Versions of Life* (Bowling Green, OH: Bowling Green State University Popular Press, 1987), 101.
5. Susan Sontag, *On Photography* (New York: Farrar, Straus & Giroux, 1977), 9.
6. Peter Osborne, *Travelling Light: Photography, Travel and Visual Culture* (Manchester, UK: Manchester University Press, 2000), 82.
7. John Urry, *The Tourist Gaze* (London: Sage, 2002).
8. Jonas Larsen, "Families Seen Sightseeing: Performativity of Tourist Photography," *Space and Culture*, 8(0) (2005): 417–434.
9. Bayard Breese Snowden, "Picture-Thinking," *Kodakery*, April 1916, 10.
10. Tim Kindberg, Mirjana Spasojevic, Rowanne Fleck, and Abigail Sellen, "The Ubiquitous Camera: An In-Depth Study of Camera Phone Use," *IEEE*

Pervasive Computing, 4(2) (2005): 42–50; Dong-Hoo Lee, "Women's Creation of Camera Phone Culture," *Fibreculture,* 6 (2005), http://journal.fibreculture.org/issue6/issue6_donghoo.html; Daisuke Okabe, "Emergent Social Practices, Situations and Relations through Everyday Camera Phone Use," presented at International Conference on Mobile Communication and Social Change in Seoul, Korea, October 18–19, 2004.

11. Kindberg et al., "The Ubiquitous Camera."

12. Nancy Van House, Marc Davis, Morgan Ames, Megan Finn, and Vijay Viswanathan, "The Uses of Personal Networked Digital Imaging: An Empirical Study of Cameraphone Photos and Sharing," presented at CHI 2005, Portland, OR, April 2–7, 2005.

13. Chalfen, *Snapshot Versions of Life.*

14. David M. Frohlich, Allan Kuchinsky, Celine Pering, Abbe Don, and Steven Ariss, "Requirements for Photoware," in *Proceedings of the ACM Conference on Computer Supported Cooperative Work* (CSCW 2002), ed. M. Twidale (New York: ACM Press, 2002), 166–75.

15. Martha Langford, *Suspended Conversations: The Afterlife of Memory in Photographic Albums* (Montreal: McGill-Queen's University Press, 2001), 20.

16. See Kris Cohen, "What Does the Photoblog Want?" *Media, Culture and Society,* 27(6) (2005): 883–901.

17. Wade Roush, "Killer Maps," *Technology Review,* 108(10) (2005): 54–60.

18. *Oxford English Dictionary,* 2005.

19. Tom Conley, *The Self-Made Map: Cartographic Writing in Early Modern France* (Minneapolis: University of Minnesota Press, 1996).

20. For instance: Harley, J. Brian, *The New Nature of Maps: Essays in the History of Cartography* (Baltimore and London: Johns Hopkins University Press, 2001).

21. John Pickles, *A History of Spaces: Cartographic Reason, Mapping and the Geo-coded World* (London: Routledge, 2004), 160–75.

22. Menno-Jan Kraak, "The Role of the Map in a Web-GIS Environment," *Journal of Geographical Systems,* 6 (2004): 83–93.

23. Joan M. Schwartz and James R. Ryan, eds., *Picturing Place: Photography and the Geographical Imagination* (London: I.B Tauris 2003); Joan M. Schwartz, "The Geography Less: Photographs and the Construction of Imaginative Geographies," *Journal of Historical Geography,* 22(1) (1996): 16–45.

24. See OECD, *Key ICT Indicators,* December, 2004, http://www.oecd.org/document/23/0,3343,en_2649_34449_33987543_1_1_1_1,00.html.

25. See National Internet Development Agency of Korea, *Survey on the Computer and Internet Usage* (August, 2007).

26. National International Development of Korea, *Survey,* 60.

27. See *The Hankyung Business,* November 28, 2004.

28. Larissa Hjorth and Heewon Kim, "Being There and Being Here: Gendered Customising of Mobile 3G Practices through a Case Study in Seoul," *Convergence,* 11(2) (2005): 50.

29. See *The Digital Times,* November 15, 2006.

30. See *The Kyeonghyang,* September 13, 2006.

31. The data was gathered from January to February in 2007.

32. This study looked through more than 400 images taken of the Cheonggye Plaza and Cheonggyechon. To understand their spatial portrayals, it examined subject matter, ways of framing, and narrative captions.

33. Lee is now president of South Korea as of the elections on 19 November 2007.

34. Myung-Rae Cho, "Can the Cheonggyechon Restoration Make Both Preservation and Development?" [in Korean], *Dangdae Byipyeong*, 26 (2004): 88–104; Seong-Tae Hong, "Neo-development and the Seoul City Government's Anti-Environpolitic" [in Korean], *Environment and Life*, 41: 52–63.
35. See Soobyung Kim, "Rather, Expose the Bottom of the Cheonggye Plaza" [in Korean], *Hangyeoryeo*, 21 (February 23, 2006): 598.

19 These Foolish Things
On Intimacy and Insignificance in Mobile Media

Kate Crawford

In *The Poetics of Space*, Gaston Bachelard recounts a letter written by Rainer Maria Rilke. Rather than selecting a letter featuring more traditionally poetic themes of love or loss, Bachelard is captivated by an account of housework. Rilke writes at length about polishing his furniture, in particular, the family piano:

> [O]ur little piano fell under my jurisdiction as duster. It was, in fact, one of the few objects that lent itself willingly to this operation and gave no sign of boredom. On the contrary, under my zealous washcloth, it suddenly started to purr mechanically . . . and it's fine, deep black surface became more and more beautiful. When you've been through this there's little you don't know![1]

Rilke details his "indispensable costume, which consisted of little washable suede gloves to protect one's dainty hands."[2] What is the value of knowing what Rilke was wearing as he cleaned his piano that day? Bachelard provokes us to find an answer to this question. While he admits "some may disdain it or wonder that it should interest anyone," he argues in defense of the writer who "braves the kind of censorship that forbids 'insignificant' confidences." Bachelard concludes that it is precisely in these revealed moments of the mundane that readers can connect their own sense of lived history with that of the writer:

> But what a joy reading is, when we recognise the importance of these insignificant things, when we can add our own personal daydreams to the 'insignificant' recollections of the author! Then insignificance becomes the sign of extreme sensitivity to the intimate means that establish an understanding between writer and reader.[3]

It is this connection between insignificant recollections and intimacy that interests me here. In particular, how the sharing of everyday actions, habits, and experiences—everyday "trivia"—forges connections between individuals who are physically remote from each other. That may be between a reader and writer, or two friends talking on the phone from different cities, or a group of thirty friends swapping messages on a microblog.

Insignificant "chatter" is a common object of complaint, particularly when it comes to mobile-phone conversations in public spaces. Similarly, exchanges by SMS text messages, in blog comments, or on social software services are often described as insignificant or trivial, and therefore pointless. Such criticisms of triviality in online and mobile-phone conversations represent a continuation of earlier criticisms of the use of the fixed-line phone, as this chapter will show. Bachelard offers a different perspective on the role of insignificant confidences: not as pointless, but as fostering a sense of connectedness by sharing the everyday. By listening more closely to apparently insignificant exchanges, we hear the gradual development of intimate understanding.

Condemnations of the quality of discussion over communications technologies abound. For example, Doris Lessing argued in her Nobel Prize acceptance speech in 2007 that the Internet "has seduced a whole generation with its inanities" and is generating what she describes as "a fragmenting culture" where good minds are lost to days of irrelevant blogging.[4] "People are increasingly documenting the most mundane and private aspects of their lives and posting them the instant they happen," writes Jennifer Saranow in the *Wall Street Journal*, creating an "assault on privacy and discretion."[5] Similar concerns are regularly aired in the popular media about the mobile phone invading public space with conversations of no importance.[6]

As researchers of mobiles have shown, such conversations and texts are less about relaying vital information and more about sharing experiential moments—which Matt Locke describes as "light touches" to intimates.[7] "Conversations about nothing" are cementing forms of social connection and intimacy, building and developing over time. Mobile media forms, by which I include mobile phones, cameras, iPods, video, and phone-networked social software, offer a variety of ways for people to share the details of daily life, sometimes with close friends and associates, but sometimes in more public, networked fora. How does the "intimacy over distance" offered by the fixed-line telephone compare with that of the mobile phone? What happens to our understanding of intimacy when sharing everyday moments occurs in public or semipublic networked space: on social sites or Flickr pages, for example? Do we need a more complex understanding of intimacy to grasp the forms of sharing and disclosing that occur through mobile media technologies?

This chapter considers the development of intimate connection in these modes where it is contained in apparently insignificant content. I use intimacy in its broad sense here, not just as romantic connection or the closeness of immediate family, but the dispersed sense of intimacy generated by regular contact with a wider circle. That may be with friends, acquaintances, and colleagues, as well as the contacts we maintain through various online and offline networks.

I do not propose that intimacy has solely positive connotations, or that it constitutes an unproblematic and unmediated form of relation. Further, intimacy in this context is not necessarily premised on sharing a literal or even honest account of daily experiences. On the contrary, what people say in person or in networked spaces such as phone calls and social networking sites may not be an accurate account of themselves. Representations may be idealized, such as when people choose to post only their most flattering photos on Flickr, or lie about their age on Facebook. Even the most mundane of intimate communications can involve a degree of deception. Indeed, outright lying is commonplace—as evidenced by a UK study in 2004 that found that 45 percent of mobile users actively lie in text messages about their whereabouts.[8]

Rather than focusing on the issue of truthfulness, my concern here is with the ways people use mobile media to connect with others by sharing everyday details: be they accurate, embellished, or entirely confabulated. Twitter provides a case study of a networked space of mobile media, as a microblogging service that can be accessed by mobile phone, instant messaging clients, or over the Web. I argue that the communicative modes of Twitter, and others like it, operate as *disclosing spaces*.[9] The "confidences" relayed in these spaces create relationships with an audience of friends and strangers, irrespective of their veracity. They build camaraderie over distance through the dynamic and ongoing practice of disclosing the everyday.

ON INSIGNIFICANCE

There is a well-established view in the popular media that mobile phones are technologies that facilitate banality in human communication. Imre Salusinszky opines in *The Australian* newspaper that mobiles not only encourage people to share insignificant details of their lives with friends, but they allow the "banal conversations of strangers" to be overheard by all nearby.[10] His view that mobiles are for "silly inconsequential" conversations is particularly targeted towards the young:

> And what, exactly, is it that keeps people, especially young adults, jabbering away into the mobiles held to their ears, or the microphones dangling from their necks? What portentous information, what profound philosophical insight, is it that can't wait until they are home again, or in the actual physical presence of their interlocutor?[11]

Salusinszky codes the mobile conversations of the young as trivial, as "jabbering" in public space. If mature adult conversation is that which is suffused with meaning and intention, then the sound of young people "just checking in" is a display of futility. This perception of young people as using the mobile phone for mere trivialities also extends to SMS texting. In an article in Scotland's *Evening News*, considerable concern was expressed about the reliance on texting by the young as a central form of communication. A psychologist is quoted as saying that despite the popularity of the medium, "to be honest, the vast majority of text messages sent seem to be trivial and irrelevant."[12]

This emphasis on the young as the carriers of irrelevant chitchat over mobiles reflects earlier attitudes about women's use of fixed-line telephones. As Lana Rakow observed in her American study *Gender on the Line*, women have commonly been criticized by men for the triviality, length, and expense of their calls.[13] In Australia, Ann Moyal conducted a similar study to investigate how women used the telephone to form and maintain social bonds, and found that regular discussions about ordinary day-to-day events were critical.[14] Moyal drew on Suzanne Keller's distinction between two genres of phone conversation: the instrumental and the intrinsic.[15] Instrumental calls, which were focused on making business arrangements, making appointments, or seeking information, were completely outnumbered in frequency and length by intrinsic calls—personal, "unpressured" exchanges, also known as idle chat.

Women's use of the phone established what Moyal describes as a "network of callers which constitute an 'electronic community' of friendships, mutual support and kin-keeping," a "'psychological neighbourhood' that substituted for face-to-face contact."[16] This intimacy over distance was maintained precisely by sharing the banalities of everyday life, by talking about what might seem to others to be insignificant details.

Thus, women and young people have both been depicted as being problematically trivial in their phone use, with their discussions of the everyday described as an inferior or degraded form of communication. Such a view gives primacy to those forms of phone use that are short, information-laden, or "instrumental" rather than intimate—characteristics which are positioned as "adult" and "male." The criticisms of women for their insignificant chats on fixed-line phones discount the emotional ties developed by these conversations, which both Rakow and Moyal identify as a form of caregiving. Likewise, the discourses about young people's trivial conversations on mobiles overlook the importance of the social bonds that are developed and maintained in these modes.

In North American media, debates about the manifest insignificance of mobile conversations are now common. As one listener remarked on National Public Radio during a show on mobile phone use:

> I overhear so many banal conversations, especially in the supermarket
> . . . I think people just talk because they can, not because they need to
> or because they have anything intelligent or important to say.[17]

The disappointment that mobile communications do not contain profound insights appears at odds with the function of such conversations—regardless of whether they are held in person, over a fixed line, or on a mobile. Much of human conversation is couched in small talk and embedded in mundane details. Then there are the exchanges that contain no substantive content whatsoever, but serve to put people at ease. As the anthropologist Bronislaw Malinowski first described when observing the Trobriand Islanders, "phatic communication"—an exchange that contains almost no substantive content—exists to strengthen social bonds, such that the "ties of union are created by the mere exchange of words."[18] In such communication, the content of the conversation is not important, but is an exchange which prioritizes a recognition of the other's presence: "How are you doing?" "I'm fine, thanks, how are you"? Such exchanges are a customary part of establishing rapport with others in person or via communications technologies.

Mobile conversations commonly contain phatic and nonphatic modes. Even when offering little by way of substantive content, exchanges about everyday occurrences represent a connected intimacy between close friends and family. As Ilpo Koskinen has noted in his study of mobile multimedia: "In our experiments, it has been a technology that 'explodes banality.' With it, people transform small things in everyday life into mutual entertainment."[19] The banal proliferates through mobile media; but it is not without function. Mobile users may be remote, but they are describing elements of their working day, their train ride, or their decisions about what to cook for dinner. They may photograph their cat, a road sign, or a particularly good cup of coffee. These moments could otherwise only be shared with domestic intimates and close colleagues, people who share working space or home space.

As a "machinery that produces banality,"[20] mobile media technologies invite remote friends and family to partake in these experiences in spite of physical distance: ties of union are maintained over distance precisely through the sharing of seemingly trivial accounts. As Ito and Okabe write, mobile phones function as a form of "glue for cementing a space of shared intimacy," between lovers, close friends, and acquaintances.[21] As fixed-line telephones offer intimacy over distance, so mobiles offer intimacy that travels with you, or "full-time intimate community."[22]

But how does this intimacy change in character when it is shared over a social network, in a public or semipublic space? Twitter, which launched in 2006, offers an example of the interplay between insignificance and intimacy in a mobile media network; it has already garnered considerable criticism (and much popularity) for trading in the banal details of people's lives.

TWITTER AND THE USES OF BANALITY

Twitter asks its users to answer the question "What are you doing now?" in 140 characters or less. Like similar microblogging services Jaiku and Pownce, Twitter is designed for brief text-based updates that are sent to friends and interested observers.[23] It could be as simple as writing about getting a coffee, or what music you are listening to, or something frustrating at work. While they can be received via the Web or an instant-messaging client, the messages are primarily designed to be sent and received by mobile phones via SMS. Twitter messages can be restricted to a private list of friends or left open for anyone to read.

One way to observe how Twitter functions is to view public messages as they appear in real time on a Google world map: http://www.twitter-vision.com. Every few seconds, new messages, or "tweets," appear from Denmark, the United States, China, Japan, Australia, the United Kingdom, France, Russia, and Iran. As one journalist writes in *The New York Times*, it is "an absorbing spectacle: a global vision of the human race's quotidian thoughts and activities."[24] Some public tweets from well-known American users include the following:

> danah boyd writes: "When oh when will my body learn to embrace mornings rather than avoid them?"

> Barack Obama writes: "Enjoying an ice cream social with residents of Sunapee, New Hampshire."

> Howard Rheingold writes: "In garden. Trying to capture essence of social bookmarking in a few minutes."

The number of Twitter users is growing rapidly. While the company does not release figures, the independent directory service TwitDir.com estimates there are around 950,000 public users who allow their profiles to be searched, while the additional number of private accounts is unknown. While it may be popular, Twitter is attracting negative descriptions that that include "pointless" and "time-consuming," as well as "problematic" and "addictive." Lev Grossman in *Time* magazine describes Twitter as the "cocaine of blogging or e-mail but refined into crack."[25] Productivity author Tim Ferriss calls Twitter "pointless email on steroids."[26] According to the science-fiction author and Twitter user Bruce Sterling, "using Twitter for literate communication is about as likely as firing up a CB radio and hearing some guy recite The Iliad."[27]

Microblogging encourages the disclosure of simple, easily described moments. Social networking researchers Ashkay Java et al. analyzed Twitter traffic and found that the greatest number of posts were about "daily routine or what people are currently doing," followed by conversations—people responding to each other's updates.[28] Twitter's emphasis on temporality (accounting for actions in the present tense) seems to exaggerate the possibility for banality. Twittering from a work computer can encourage messages about the office, or discoveries on the Web, while message from a phone can encourage "out in the field" reports from events, dinners, gigs, or trains.

According to Evan Williams, founder of Obvious, the company that developed Twitter (and Blogger), the three main criticisms of the service persist: "Why would anyone want to do this?" "It's pointless," "It's trivial."[29]

But it is this very mundanity that is central to Twitter's success. As a service, it offers us access to the everyday thoughts of people we are interested in. Rather than the more substantial writing that may be developed in blogs, "tweets" record the moments that are not usually saved for posterity, brief moments that normally disappear beneath the surface of life. In Maurice Blanchot's words, "the everyday escapes, it belongs to insignificance."[30]

So why is it that groups of friends, associates, and strangers delight in reading these insignificant details from the lives of others? In Bachelard's view, it is precisely through the joy of recognizing the importance of insignificant things that writers and readers connect, and there is a tightly wound loop between the roles of reading and writing on Twitter. Users switch from being one to the other in the space of a moment.

The cofounder of Twitter, Jack Dorsey, has described how people respond when they first hear that tweets are mainly about simple moments such as cleaning the bathroom or boiling the kettle:

> The first reaction is to hate it because it's seen as the most useless thing in the world and no one would ever want to know about boiling water. But these small details in life are what connect us most. Everyone has these specific moments and you normally don't bring them up in conversation because it seems so trivial but it's not, it's really important.[31]

Dorsey's belief in the importance of small details begs the question of why such moments in life are of consequence. How does the sharing of daily actions and thoughts operate to connect people and create intimacy? Elspeth Probyn raises this question in the essay "Thinking Habits and the Ordering of Life."[32] Probyn begins with the example of cleaning her floors:

> I mop the floor all the while chasing the notion of semiotic habit around the dusty corners of my mind. The floor looks good, a legacy of having been a commercial cleaner in my youth.[33]

Probyn's domestic duties perform a double function. She cleans her floors, but she also tells us about cleaning her floors, and what thoughts cross her mind in the act. In these embedded domestic moments, she seeks to move beyond the division described by Charles Pierce between "the outer world" of social reality and the "inner world" of subjectivity.[34] The fabricated nature of this dividing line is revealed by paying heed to ordinary tasks and mundane details.

In a similar way, Twitter can be understood as a mechanism that commingles these two worlds, where "the inner and the outer continually move through each other."[35] It is the sharing of subjectivity, as uneventful and routine as it may be, that forms the basis of Twitter's social reality.

Cleaning, in its many forms, is a popular topic on Twitter. For example, Jim Wimpey writes: "Cleaning up post-party house. Intermittently returning to computer to read iPhone news." Good Egg333 writes: "If the question is what is the best cleaning aid in the world, the answer is a gin martini." Evo_terra writes: "Spent the last hour cleaning up my LinkedIn profile. 'Tis now bright, shiny and new!" These are just some of the types of cleaning mentioned on Twitter: in domestic space, office space, but also the many kinds of electronic self-maintenance that are conducted every day. It may be organizing an e-mail in box, deleting texts on a phone, updating profiles on MySpace, LinkedIn or Facebook. People are regularly Twittering on these topics, as the answer to "what are you doing now?" is often something as banal as cleaning and sorting, in all its forms.

Feminist theorists have struggled with the importance of the trivial tasks of the domestic everyday for many decades. Sylvia Bovenschen wrote in 1976 about the limitations of the ordinary tasks of household and self-maintenance:

> [These activities] remained bound to everyday life, feeble attempts to make this sphere more aesthetically pleasing . . . But [housework] could never leave the realm in which it came into being, it remained tied to the household, it could never break loose and initiate communication.[36]

Twitter is a networked space where we can see the breaking loose of these everyday acts, as they become the basis for communication. Insignificant details, the ordinary, the domestic are the ties that bind groups of users together. Further, Twitter offers different analytic perspectives on how people enact intimacy and connection. It reverses the idea of isolated users receiving thin channels of human contact, where trivial details or chatter are deemed to be empty forms of communication. Instead, small details and daily events cumulate over time to give a sense of the rhythms and flows of another's life. The background awareness of others offered by Twitter has been described as "social proprioception": a subtle, shifting knowledge of where people are, what pressures or pleasures they are experiencing.[37] Leisa Reichelt calls it "ambient intimacy": an "ongoing noise" of the everyday experiences of people one cares about.[38] Beyond the restricted understanding of the intimate that prioritizes exchanges of gravity and magnitude, Twitter represents something more molecular and dispersed.

However, Twitter's capacity to connect people via short reports of their activities can also generate forms of claustrophobia and distaste. Twitter updates can literally interrupt one's working day—particularly as some Twitter IM clients "pop up" recent tweets from friends directly on screen. These messages can provide a moment of respite or amusement, or they can be an unwelcome disruption. Some users only access Twitter over the Web, with no mobile or instant message access, in order to actively choose when to receive messages from friends rather than live with such an ongoing presence.

In addition to the potential invasiveness of sharing subjectivity, some users criticize the genre of messages that commonly appear on Twitter, particularly if they had a work-related emphasis. Melissa Gregg describes in her blog why she stopped using Twitter:

> [T]he group of fellow Tweeters I accumulated skewed the updates I received in certain directions. Many being work colleagues and/or geeky types, I would know what conference someone was speaking at or the latest software they were frustrated with but I wouldn't know if they had been particularly moved by something or what they wanted their life to be like in 3 months' time.[39]

In Gregg's view, the blurring of "friends" with work-related "contacts" in her Twitter profile resulted in a slanted form of intimacy. Others, however, have commented that they enjoy the "disjointed conversations" of Twitter while working from home, precisely because it functions "like a water-cooler or lunch room" in a shared office environment.[40] For some, Twitter provides a valuable form of interaction with a dispersed social space.

But the composition of that social space can present problems. Users ultimately must decide how to construct their own environment in Twitter (a public or private profile; a set of friends with or without work "contacts"); however, there are few established forms of etiquette when it comes to turning down friend invitations. If these environments operate as a form of networked intimacy that includes close friends, family, distant acquaintances, colleagues, and strangers, the negotiations about how much and which elements of the "inner world" to share are delicate.

In addition, the conversational field presented by a Twitter network can be fractured in two critical ways: being disjointed in time and reaching different publics. First, messages may not be seen and responded to until after a posting has lost its currency. Second, messages are broadcast to a user's entire contact base, but when friends' social networks do not exactly coincide with one another, each interlocutor is effectively speaking to a different audience. Conversations can thus seem out of joint: a message is answered too late, or a "crossed line" effect emerges when people have lost the thread of a discussion.

But beyond these issues, there are varied processes by which users craft and curate their own presence on Twitter. While many people post about their housework or their lunch, some use microblogging as a chance to write haiku-style fiction.[41] Others manage sharp satire in 140 characters or less.[42] A few have become minor celebrities, such as Merlin Mann, the founder of productivity site 43 Folders and one of the most followed people on Twitter. Most users fall somewhere between these styles, at times reporting the everyday, or throwing in amusing one-liners or creating fictional "stories."

The affinity created in these less directly personal modes is one of *voice*: becoming accustomed to the style of someone's humor or flights of imagination. For example, Merlin Mann posts about a fictional breakfast with Ayn Rand: "Patched things up over a cordial breakfast with Ayn Rand. My eggs were fine, but she declared the oatmeal 'weak' and 'unheroic.'" Mann's fabricated meals with the rich and famous then become a running thread, a riff developed by him and then expanded by those amongst his 10,000 Twitter followers. This is the intimacy of the in-joke.

Thus, while the general operation of Twitter is the gradual accretion of everyday moments and passing thoughts, there is considerable variation in how people adapt this process. Some use it to describe what they are doing; others use it to share information or converse; others confabulate and entertain. Regardless of the veracity or banality of these contributions, they are read by a community of users who come to recognize and relate to that presence, tracking their moods, habits, and whims. In this way, Twitter is best understood as a disclosing space, with all that entails: truth, falsity, humor, triviality, drudgery, gossip, and camaraderie.

ONE-TO-ONE INTIMACY OR A NETWORK OF DISCLOSING SPACES?

As this chapter has shown, the discourses that oppose insignificant conversation (women chatting over the household phone line, young people "jabbering" into their mobiles on the train) are discounting the forms of care work, bonding, and intimacy that are created and maintained in such modes. An apparently trivial exchange is a single scene in a long enactment of being social that incorporates many moments of solitude, connectedness with one, and connectedness with many.

Writing about the experience of a face-to-face discussion, Erving Goffman argues:

> [T]alk is unique, however, for talk creates for the participant a world and a reality that has other participants in it . . . We must also see that a conversation has a life of its own and makes demands on its own behalf. It is a little social system with its own boundary-maintaining tendencies; it is a little patch of commitment and loyalty with its own heroes and its own villains.[43]

Goffman's work has been taken up by mobiles scholars who are seeking to understand how ritualized communication is used to develop social bonds.[44] While the possible channels for "talk" have expanded since Goffman published *Interaction Ritual* in 1967, conversations continue to build affective ties and define boundaries regardless of the medium in which they occur. As Hjorth has noted, the practice of intimacy has always been mediated;[45] so we can find many commonalities between in-person, fixed-line telephone, mobile, and socially networked conversations. But there are also important differences.

Although the importance of sharing the everyday remains, the available technologies to do so evolve and create new spaces, subtly changing the capacities of emotional exchange. As the fixed-line telephone offered the intimacy of listening over distance, so the mobile phone allows those conversations to happen in new territories: textual, visual, and aural, moving along with the user. Mobile social networks offer us yet more angles of perception on human rapport and its plasticity. In semipublic networks such as Twitter, new collective intimacies develop—social groups that bond through the minutiae of their lives, gradually developing a more granular awareness of each other.

Each medium also presents a set of limitations. Twitter's restriction to 140 characters per message provides a clear example of a designated boundary that preconditions the kinds of communication that are possible. Its emphasis on reporting the quotidian also constitutes a restraint on other forms of discourse. As we have seen, Twitter users find many ways to adapt the form to their own ends, but there is a limit point: such a space is not conducive to lengthy political debate or detailed analysis. As Bovenschen reminds us, the everyday also creates a bind. While it is valuable and important in human connection, can it ever transcend its realm to instigate other forms of debate, reaction, or change? Cultural and feminist theories about the everyday offer us tools to reframe the value and function of "idle chat" over communications technologies, and while they point to the structural limitations, they also suggest its crucial role in building human intimacy.

Sustaining multiple conversations at once is a commonplace—a discussion over a fixed line at work while responding to e-mail, replying to instant messages while chatting across a kitchen table, sending a tweet while listening to a band at a pub. These are all different kinds of disclosing spaces, some being one-on-one, others being within a closed group of friends, or open to whoever is listening. Disclosing requires a listener in order to constitute a "disclosure," and as the technological modes of speaking have changed, so have the modes of listening. As mobile-media forms offer us new spaces of disclosure, so we develop capacities for hearing in different ways: face to face, over the phone, or just "in the background" as we listen to channels of personal daydreams and insignificant chatter.

While Goffman has emphasized the importance of "talk" to social bonding, I argue it is these developing modes of "listening" that underscore what is particular to intimacy in networked mobile media. Twitter is one instance of a space where "listening in" to the disclosures of others occurs continuously, whether one is talking back or not. It is this diffuse familiarity, more indirect than a phone call or a personal e-mail, that points a way forward to conceptualizing these emerging spaces of mobile media. As Bachelard suggests, tuning in to the detail of everyday life reveals the many, fine-grained ways that intimacy is established and sustained.

NOTES

1. Gaston Bachelard, *The Poetics of Space*, trans. Maria Jolas (Boston: Beacon Press, 1994), 70.
2. Bachelard, *Poetics of Space*, 70.
3. Bachelard, *Poetics of Space*, 71.
4. See Maev Kennedy, "Nobel Prize Winner Lessing Warns against Inane Internet," *The Guardian*, December 8, 2007, 20.
5. Jennifer Saranow, "The Minutes of Our Lives—Small, Private Moments Get Live Blog Treatment," *The Wall Street Journal*, March 2, 2007, W1.
6. See Sadie Plant, *On the Mobile: The Effects of Mobile Telephones on Social and Individual Life*, report for Motorola, 2002, http://www.motorola.com/mot/documents/0,1028,333,00.pdf; in media debates, see Patrick Begley, "Victims of Inane Conversations," *The Courier Mail*, December 19, 2007, 37.
7. Matt Locke quoted in Larissa Hjorth, "Locating Mobility: Practices of Co-presence and The Persistence of the Postal Metaphor in SMS/ MMS Mobile Phone Customization in Melbourne," *Fibreculture Journal*, 6 (2006), http://journal.fibreculture.org/issue6/issue6_hjorth.html.
8. Will Sturgeon, "British Are a Nation of Cowardly Texters," *Silicon.com*, January 21, 2004, http://networks.silicon.com/mobile; see also Genevieve Bell, "Secrets, Lies and Digital Deceptions," LIFT08 conference presentation, February 7, 2008, http://www.liftconference.com/.
9. I am adapting the phrase *disclosing spaces* from art historian Andrew Benjamin, who uses the term to describe paintings and their intertwined relationship with criticism. Andrew Benjamin, *Disclosing Spaces: On Painting* (Manchester. UK: Clinamen Press, 2004).
10. Imre Salusinszky, "Lost in Conversation," *The Australian*, April 15, 2007, 5.
11. Salusinszky, "Lost in Conversation."
12. See Sandra Dick, "Luv It or h8 It, It's Here to Stay," *Evening News*, January 25, 2005, 16.
13. See Lana Rakow, *Gender on the Line: Women, the Telephone, and Community Life* (Urbana, Champaign, IL: University of Illinois Press, 1992).
14. Ann Moyal, "The Feminine Culture of the Telephone: People, Patterns and Policy," *Prometheus*, 7(1) (1989): 5–31.
15. See Suzanne Keller, "The Telephone in New (and Old) Communities," in *The Social Impact of the Telephone*, ed. Ithiel de Sola Pool (Cambridge, MA: MIT Press, 1977), 281–98.
16. Moyal, "Feminine Culture of the Telephone," 6.
17. Neil Conan, "Cell phone Technology and Etiquette," *NPR: Talk of the Nation*, 14 July, 2005.
18. Bronislaw Malinowski, *Argonauts of the Western Pacific: An Account of Native Enterprise and Adventure in the Archipelago of Melanesian New Guinea* (London: Routledge, 1922), 315.
19. Ilpo Koskinen, "The First Steps of Mobile Multimedia: Towards an Explosion of Banality," paper given at *The First Asia–Europe Conference on Computer Mediated Interactive Communications Technology*, Tagaytay City, The Philippines, 20–22 October, 2003, pp. 19–20, http://www2.uiah.fi/~ikoskine/recentpapers/mobile_multimedia/manila-koskinen.pdf.
20. Ilpo Koskinen, *Mobile Multimedia in Action* (New Brunswick, NJ, and London: Transaction Publishers, 2007), 12.
21. See Mizuko Ito and Daisuke Okabe, "Technosocial Situations: Emergent Structurings of Mobile Email Use," in *Personal, Portable, Pedestrian: Mobile Phones in Japanese Life*, ed. Mizuko Ito, Daisuke Okabe, and Misa Matsuda (Cambridge, MA: MIT Press, 2005).

22. See Ichiro Nakajima, Keiichi Himeno, and Hiroaki Yoshii, "Ido-denwa Riyou no Fukyuu to sono Shakaiteki Imi" [Diffusion of Cellular Phones and PHS and their Social Meaning], *Tsuushin Gakkai-shi* [*Journal of Information and Communication Research*], 16(3) (1999), and Misa Matsuda, "Introduction: Discourses of Keitai in Japan," in *Personal, Portable, Pedestrian: Mobile Phones in Japanese Life*, ed. Mitzuko Ito, Daisuke Okabe, and Misa Matsuda (Cambridge, MA: MIT Press, 2005).

23. See http://www.twitter.com, http://www.jaiku.com and http://pownce.com.

24. Jason Pontin, "From Many Tweets, One Loud Voice on the Internet," *The New York Times*, April 22, 2007, 3.

25. Lev Grossman, "The Hyperconnected," *Time* (Canadian edition), April 16, 2007, 40.

26. Quoted in Clive Thompson, "How Twitter Creates a Social Sixth Sense," *Wired* magazine, June 26, 2007, http://www.wired.com/techbiz/media/magazine/15-07/st_thompson.

27. Bruce Sterling quoted in Jason Pontin, "From Many Tweets, One Loud Voice on the Internet," *New York Times*, April 22, 2007, 3.

28. Ashkay Java, Xiaodan Song, Tim Finin, and Belle Tseng, " Why We Twitter: Understand the Microblogging Effect in User Intentions and Communities," proceedings of *WebKDD/SNAKDD 2007: Workshop on Web Mining and Social Network Analysis*, August 12, 2007, 56–66.

29. Evan Williams quoted in Kate Greene, "What Is He Doing?", *MIT Technology Review*, November 1, 2007, http://www.technologyreview.com. The name of Williams's company, Obvious, is also a synonym for banal, the term which has regularly been used to criticize both of Williams's major products: Blogger and Twitter.

30. Blanchot quoted in Rita Felski, *Doing Time: Feminist Theory and Postmodern Culture* (New York: New York University Press, 2000), 78.

31. Jack Dorsey quoted in AFP, "Net's a Twitter with the Sound of Mundane," *The Australian*, June 12, 2007, 27.

32. Elspeth Probyn, "Thinking Habits and the Ordering of Life," in *Ordinary Lifestyles: Popular Media, Consumption and Taste,* ed. David Bell and Joanne Hollows (Milton Keynes, UK: Open University Press, 2005), 243–54.

33. Probyn, "Thinking Habits," 246.

34. Pierce quoted in Probyn, "Thinking Habits," 253.

35. Probyn, "Thinking Habits," 254.

36. Bovenschen quoted in Probyn, "Thinking Habits," 252.

37. Clive Thompson, "How Twitter Creates a Social Sixth Sense," *Wired* magazine, June 26, 2007, http://www.wired.com/techbiz/media/magazine/15-07/st_thompson.

38. See Leisa Reichelt, "Ambient Intimacy," Disambiguity Blog, posted on March 1, 2007, http://www.disambiguity.com/ambient-intimacy.

39. Melissa Gregg, "What I Am and Am Not Doing," *Home Cooked Theory*, posted on September 10, 2007, http://homecookedtheory.com/archives/2007/09/10/what-i-am-and-am-not-doing.

40. Glenda Hyatt, comment on "I love Twitter, but I Have to Quit It," Jim Kukral Blog, comment posted November 28, 2007, http://www.jimkukral.com/i-love-twitter-but-i-have-to-quit-it/. Hyatt has cerebral palsy, but uses Twitter to stay in touch with others while she works from home as a writer and blogger.

41. See, for example, 140 Story: http://www.twitter.com/140story.

42. There are many possible examples here, with two being the American user LonelySandwich and the Australian user RosieFantail.

43. Erving Goffman, *Interaction Ritual: Essays on Face-to-Face Behavior* (New Brunswick, NJ: Transaction Publishers, 2005), 113–14.
44. See, for example, Rich Ling, *New Tech, New Ties: How Mobile Communication Is Reshaping Social Cohesion* (Cambridge, MA: MIT Press, 2008).
45. Hjorth, "Locating Mobility."

20 Mobility, Memory, and Identity

Nicola Green

. . . the struggle of man against power is the struggle of memory against forgetting.[1]

Ours truly is a present that is flooded by future as well as past possibilities.[2]

INTRODUCTION

The study of memory in the "Western" world has often been seen as the purview of social historians, studying the ways in which artifacts of the recent (and more remote) past have informed the ways in which societies and cultures reproduce themselves, and change over time. As such, its historiography has encompassed questions about the cultural and political impetus (and consequences) of individual, and collective, remembering and forgetting. Sociology's part in the study of memory has been more ambivalent and uneven, hovering on the outskirts of these debates. On the one hand, interactionist sociological traditions have historically developed accounts of social memory that focus on narrative auto/biography, and the interactive and intersubjective sociality of memory-making. On the other hand, critical historical sociology has provided accounts of the politics of collective and ritualized memory practices in the Western world. It is only more recently that a renewed interest in memory has reemphasized the importance of the temporal dimensions of human life, and how processes of remembering, forgetting, processing, and selecting the artifacts and experiences of the rhythms of life, connect both the past, and the future, with the social present.[3]

One reason, perhaps, for this renewed interest, is the recognition that the technologies and artifacts through which memories are constructed are rapidly changing with new media, digitization, and mobility. At the same time, particularly in the intersection of cultural theory, globalization and changes in time/space, the sociological consideration of imaginaries of memory (and memory work) as central activities in (post) modern, mobile, and mediated societies has become more and more central. For their part, various cultural theorists of "postmodernity" have pointed to "cultural amnesia"[4] as a predominant feature of contemporary life, resulting from the compression of time and space, and acceleration in mobility, in a more transient, ephemeral, and malleable "liquid modernity."[5]

Media technologies and practices of mediation have long been considered central to subjective, sociocultural, and institutional forms of remembering, memorializing, and forgetting, and it is increasingly recognized that new *mobile* media are again intervening in memory practices in diverse ways. This is reflected in both the growing attention paid to mobile media of memory in the computing and commercial domains, and by a recognition of shifting cultural patterns of memory-making in changing times. What the activities and arguments regarding digital and mobile memory call most sharply into focus (again) is that memory is not an unchanging snapshot of the past, but rather, memory practices rather change constantly with life and values of particular groups—including the ways those shift and change with new technologies and mediated patterns of living. As Van Dijck notes:

> Digitization is surreptitiously shaping our acts of cultural memory—the way we record, save and retrieve remembrances of our lives past. With the emergence of every new technology, from print to photography and from the gramophone to the computer, people have hailed and despaired new means for (self-)recording, storage and retrieval . . . [in] "mediated memories": memories recorded by and (re)collected through media technologies. . . . Few scholars have bothered to theorize the shaping power of media technologies on the materialization of cultural memory—a shaping power that is particularly discernible in periods of technological transformation.[6]

This paper aims to explore the ways in which increasingly *mobile* media are intervening in the cultural and political economies of memory practices. The paper aims to consider the increasingly interwoven *scales* of memory-making in mobile media in particular—from the autobiographical and intersubjective to the collective, the institutional, and the archival. On the one hand, the sociocultural practices of memory-making are diversifying, and the terms of their materiality, the artifacts that both produce and are produced by memory-making practices, are changing in their production, distribution, and consumption. The acquisition of both sociocultural and economic *value* makes these memory-making practices part of a cultural logic beyond their affective and interrelational values. According to Packard, ". . . electronic, digital, and biological technology provide more and more 'forms' of memory beyond social, collective, or traumatic memories . . . patented and sold as product (. . . *commodity memory*)."[7] On the one hand, while technically mediated memory work can produce both proliferating digital artifacts as well as enabling the "instant" forgetting of the digital form. On the other hand, the archival impulse seems to be spreading, particularly at institutional scales, producing increasingly commodified "memories" about individuals and populations over which they have no control. The logics of data retention, and who controls what is done with it, intervene in the power relations of memory-making, and how both institutional and commodity memories are shaped and used. Against this backdrop of the cultural and political economies of digital media in general, the chapter addresses the specificities of mobile media in particular—mobility highlights the performativity of mediated memory-making, the ambiguity surrounding the materialization of memory artifacts, and the tensions between the collection of memory artifacts, how they are held and distributed, and therefore the ways in which practices of remembering and forgetting are being transformed.

MEMENTOS OF (POST)MODERN IMAGINARIES

My starting point for this discussion is in popular cultural representations of those millennial desires and anxieties surrounding memory and remembering that accompany contemporary transitions in media technology and cultural practice. Consider, for example, the film *Memento*[8] wherein the central character, Leonard Shelby, experiences "anterograde memory dysfunction"—the ongoing loss of all short-term memory after trauma. The film gives us a sense of the specific ways that memory and media technologies are intertwined—in the construction of a coherent narrative of self, in the construction of intersubjective relations with others, and in the construction of collective and institutional narratives of life.

In *Memento*, Leonard Shelby, the central character, experiences "anterograde memory dysfunction" after witnessing the rape and murder of his wife—the ongoing loss of all memory, on an ongoing basis, after the traumatic event—he can't "make new memories." The premise of the film is that he therefore forgets—constantly, from the moment of trauma. And yet still he embarks upon a project to reestablish his memories of life after his wife's murder, to concretize events, and to *remember* in forms *other* than as expressed by his conscious, waking thought processes. The plot and narrative of *Memento* therefore works backwards—in mimicry of the dysfunction of the central character himself. Each moment, event, dialogue works backward from a *diffuse* and *unstable* point of present consciousness and subjectivity, *to* an (albeit imaginary) originary point, the moment at which his wife is murdered, and his memories leave him. In the film, Leonard Shelby remembers his name, aspects of his self, and his life before the traumatic events that now affect his memory. However, the film narrative further demonstrates the ways that memory *technologies* intervene in the active (re)construction of memories, through the recovery of *loss*, at different scales of social relationality.

It is this attempted recovery of loss—the particular "struggle of memory against forgetting" in which Leonard is engaged—that expresses the contemporary anxieties surrounding memory that are reflected in the film. For Leonard, as for all of us, memory is central to his autobiography of self and subjectivity, to the construction and reconstruction of expressed narratives of the self via subjective processes. Leonard Shelby must consciously and constantly recon-struct a narrative of his life in order to attempt to reestablish his subjectivity as a functional human being. He says, simply, "I want my life back," as even if *alive*, he has no *life* without memory. One must remember or forget, whether consciously or unconsciously, where one has been, to be where one is, and where one is going, in a narrative of life (however constantly constructed this is, as in Leonard's case). This is, of course, entirely consistent with the ways that social theory has conceptualized the ways that subjective self-identity is constructed from multiple fragments of memory and experience, which are narrativized—and revised, and re-presented, depending upon context—over time. The story of Leonard Shelby simply makes this process explicit.

Social theory has also pointed out that even in the case of self-identity—of subjective narratives of self to the self—memory and identity are, secondly, *also always* intersubjective and relational. Narratives of self emerge from the processes of constructing relationalities with others on a day-to-day basis with respect to past and present events, and shifting understandings and inter-pretations of them. In every event and dialogue in the film, Leonard Shelby's self is re-created through a shifting relationship with every other character, within their collective understandings of how autobiography "works," and how the self is lived and narrativized. As Jeffrey Stepinsky notes,

> As a socially mediated expression, memory emerges in an act of in-terpersonal negotiation and meaning making. Furthermore, the social construction of memories is grounded by the forms of mediation avail-able to a culture in a particular time and place. Communities, then, cannot remember in any particular manner—it is not a kind of social construction that is radically open and relative . . . Rather, the way that one remembers and the contents of memory must resonate with larger cultural and social patterns . . . [which] is to say that human collectives enter into a kind of dialogue with the conditions of their time to clarify and establish a self-understanding . . .[9]

Our interpersonal, social memories, therefore, intersect with scales of memory at the collective and institutional levels, anchored via various tech-nologies of mediation. It is these technologies of mediation that are one of the most compelling features of *Memento*: those—mobile—technologies Leonard uses to attempt to effect his reconnection with the world, the *forms of mediation* available to him, and which provide the social, cultural, and interpersonal resources with which to re-narrativise his life. These media-tion technologies do *memory work*—they are Leonard's recording devices and activities, from text in the form of written notes on paper or on objects such as receipts or beer mats, to visual and photographic images of objects, people and places, to lists tattooed on his own body. *Memento* therefore demonstrates how memories become *embodied* through technologized forms of mediation—become part of "habitus," "techniques of the body," and "techniques of the self"[10]—not only in ritual commemoration and memorializing[11] but also in the course of everyday life.

Our prosthetics for memory are sometimes similar to those of Leonard, sometimes different. Leonard carries with him a whole range of mobile and mediated technological objects with which to materialize, concretize, his narrative of self—as, so often now, do we. I suspect that most of us would opt for the use of a PDA (personal digital assistant) to upload a blog (or a mobile phone or digital camera to preserve an image) rather than tattooing a list on our own body, but the principles are not dissimilar. In doing so there are ways in which we share Leonard's anxiety both to *remember*, to narrativize a self in the face of "amnesia," as well as to *document*, to externalize memory within the sphere of conscious control. In Van Dijck's terms, these anxieties have become pervasive in Western experience economies:

> For an upscale western audience, managing data has become an attractive metaphor for controlling life. To live an experience at a date and time of one's choosing—rather like a television programme recorded on a VCR—takes some pressure off life's fast pace, regulated by the clock. What could be more appealing to a contemporary user, struggling with time constraints in an experience economy, than the storage of events in mediatized, retrievable memories? The anxiety of missing one's children growing up can be assuaged by the thought of a personal memory machine, enabling precious moments to be replayed at a time more convenient than the ever-demanding present. Experiences etched in dimensions of time hence become a timeless repository of reruns. On the other end of the spectrum, the prospect of harrowing memory disorders, such as Alzheimer's, feeds on another anxiety: the anxiety for amnesia. Complete storage of personal memory and collections should avoid the erasure of someone's unique identity. The anxiety of forgetting is intertwined inextricably with the anxiety not to be forgotten . . .[12]

The anxiety for amnesia correlates to the impulse to the *exteriorization* of subjective memory as lived—the impulse to concretize fragments of time and subjective impressions, and to do so in ways that both promote sociality and contribute towards collective remembering, within the structures of experiencing and knowing available to our technologized culture and forms of mediation. Our "new" mobile media intervene in and reconfigure the construction of social and cultural identity through memory work similar to that which Leonard carries out. Mobile phones, PDAs, and their increasingly networked connections via the Internet can be central as a space of memory, from the symbolic as well as practical purposes of address books, the activities of not only sending but keeping texts (as "keepsakes"), or the activities of sending digital photos amongst groups of family and friends, which are also then kept, passed on, and shared. As with the case of Web pages, e-mail, and MSN conversations, these *social remembering* practices are transforming the memory practices of the past.

The exteriorization of memory has been claimed as a particular feature of modernity, a particular product first of the technologies of writing (as distinct from oral cultures in which memory-making is embodied more firmly in ritual narrative practices), and later in the rationalization and bureaucratization of social life within institutions, and the development of further technologies to fix traces of the past in the present, such as photography and computing. According to Paolo Jedlowski, modernity has produced, on the one hand, "a world in perpetual change, in which traditions lose their value and recurring discontinuities are generated; on the other hand it has offered ever more sophisticated technical instruments that exteriorize the human faculty of recall and question its meaning."[13] The "modern" world requires us increasingly to concentrate on and in the present, even as the objects and sense prosthetics that assist our recall proliferate and multiply.

Both the technical and cultural trajectories intensify and converge in what some have deemed the "post"-modern. Dealing with broad themes of the transformation of time and history in the contemporary world, a number of scholars have pointed to the intensification of time relations with respect to historical events, media representation, and memory.[14] The thesis that the mediated representation collapses an event, its history and context, and its representation, is now familiar and articulated more fully elsewhere. For my purposes here, however, Andreas Huyssen encapsulates the "cultural amnesia" to emerge from this collapse by arguing that:

> A sense of historical continuity or, for that matter, discontinuity, both of which depend on a before and an after, gives way to the simultaneity of all times and spaces . . . But this simultaneity and presentness, suggested by the immediacy of images, is of course largely imaginary, and it creates its own fantasies of omnipotence: channel-flicking as the contemporary strategy of narcissistic derealization. As such simultaneity wipes out the alterity of the past and present, here and there, it tends to lose its anchor in referentiality, in the real, and the present falls victim to its magical power of simulation and image projection.[15]

For Andrew Hoskins, commenting on the implications of mass-broadcast television for the construction of memory, these images produced "represent an inexhaustible 'external' memory . . . that collapses memory simultaneously into the present (the only space the televisual image inhabits) and into the archive . . ."[16]

Couple the cultural logics of speed, simultaneity, and instantaneity with the increasing mobility of people, technologies, and images, as well as a global data network, and a situation is produced wherein none of the reproduced and re-presented images, texts, mediations act as an anchoring point in any time or space, or to any particular relational configuration.

The "cultural amnesia" described implies that no collective narratives can now serve as expressions of identity or common "points of reference" in the context of mediated globalization.[17] Instead, as John Lury remarks, when coupled with interactive media, such speed-induced representational artifacts constitute "the combination of the exhaustive retrievability and inability to remember that is random access memory."[18] In a context of "random-access" memory, the desires and anxieties associated with memory and its loss are often cast as deep uncertainties about practices of remembering—not about what *should* be remembered, but what *could* be remembered, given technological possibilities.[19] It is at this point that technologically mediated and mobile memory practices connect directly with cultural logics to configure new patterns of remembering and forgetting.

SCALES OF MEMORY AND THE SOCIAL COLLECTIVE

Against the desires and anxieties articulated by Leonard Shelby's story, and the specter of "cultural amnesia," what then is the role of specifically *mobile* media in memory-making practices? What are their connections with existing media formations, and what roles are they playing in ever-more-convergent nodes of representation? What impacts are felt at different scales of memory where mobility becomes part of the memory-making framework?

Some argue that "mobility" is itself one factor contributing to a progressive alienation of the self in modernity, and thus to the "cultural amnesia" described earlier. Celia Grech notes, following Grossberg, that the logics of modernity fragment identifications of self and other, and of self to the self, a process both made possible through modern technologies of mobility and intensifying with the convergences of new media and increasing speed:

> If other-ing is one facet of alienation, and alienation one facet of modernity, this helps explain the intense interest in nomadology, migration, exile, and mobility as modern thinkers try to understand the impact of modernization. And, if the engine of modernity is capitalism (the grand economic theory that envelops the world today) and the technologies (modes of production) it engenders, isn't new media complicit in further alienating us?[20]

As Hoskins notes, however, the past is not necessarily "lost" in an instantaneous present (in which selves are necessarily "alienated"), but is instead *transformed* with respect to both the present and the future, as well as the media and memory.[21] Leonard Shelby's story is instructive with respect to what some of those transformations might be in the realm of mobile media—*Memento* makes explicit the ways that mobile technologies can reconfigure the construction of autobiographical, sociocultural, collective, and institutional memory, through a variety of technologies.

In *Memento*, Leonard's autobiographical and subjective story of *himself*—his conscious actions and interactions—is constructed through lists, notes on paper, posters glued to mirrors, or tattoos on his body. Contemporary mobile technologies instead comprise different kinds of personal (and personalized) prosthetics. Mobile phones, PDAs, digital cameras, and their increasingly networked connections via the Internet are all means to construct *archives of the self* as central spaces of autobiographical memory, from the symbolic as well as practical purposes of mobile phone address books (lists), the activities of not only sending but keeping texts (as "keepsakes"), or the activities of sending digital photos from digital "cameras" (of various sorts) amongst groups of family or friends, which are also then kept, passed on and shared between the (mobile) laptops that sort, collate, and archive them.

Increasingly, "moblogs" or "mobile blogs" can fulfill similar functions, as a chronological and textual record of both the subjective and relational self. The textually based journals or diaries of the past were initially supplanted by (or were accomplished as complementary to) personal Web pages. The combination of personal diary and Web pages spawned the "weblog" or "blog," enabling the proliferation of multimediated spaces to describe, explore, and, indeed, *display* the "self." Although similarly multimedia and networked, the "moblog" represents the opportunity for the continuous archiving of subjective experience moment by moment, throughout an increasingly mobile existence.

In all of these contemporary examples, mobility is one of the key relations. The technologies that promote the *self-archive* "on the move" (and convergent with further fixed-line and database technologies) make explicit the ways in which the context of the autobiographical is always in the relational. Hawlbachs established that any subjective memory is social in nature, created in and maintained by the interactional and collective—but the balance between these aspects of memory shifts with respect to the relative productive power of each element in the case of digitization and mobility. In Leonard Shelby's case, his narratives of self and their technological mediation are paralleled by stories of, and by, others—notes from his friend Teddy, photographs of the people he encounters, and events in the story are framed through the sharing of notes, images, events and stories with others, for mutual consideration. In our own practices, accounts and archives of self are simultaneously narratives of others and *social archives*. They are social in at least two senses—in that they provide one possible account of a network or group, and that they are made, remade, interpreted, and reconsidered in association with others.

Commercial companies and computing projects have paid some attention to autobiographical digital memory technologies, researching and designing products and services to replace the analogue practices of, for example, photo albums or photo frames, or systems of self-archiving in "shoeboxes." As digital initiatives, however, they tend to focus on the digital archive as the account of self (an archive of life experience that provides a total and comprehensive database to be called up in the assistance of recall), and tend to neglect the interactional aspects of the simultaneously social archive. While on the one hand these practices provide an account of a life, on the other hand they also provide opportunities for sociality—to be together and to share moments, to recall and reinterpret, to mutually shape the past (and therefore present). Because they tend to focus on the house and the household (as well as the individual) as locales of memory, they also neglect the performative as well as interactional aspects introduced by mobility (collectively viewing texts and images on a mobile phone while on the move as well as sending them to other mobile devices, or the posting of an account of everyday travels through a city while conducting them). When archiving practices are dispersed in time and space, and interaction may happen at a distance as well as in copresence, the performativity of a self-archive is conducted through the networked digital artifacts that are in constant circulation between both distant and copresent others, via numerous devices and networks—the social archive.

The connections mobile media make to other networked technologies, and the social archives these create, can therefore also make more explicit that age-old dilemma of modernity—the relationship between what is private and what is public in collective memory. In Western modernity, the archives of group memories (often cast as the collective memories of a nation-state) were mainly historical and considered public institutions— examples being museums, and more explicitly articulated "memorials" to historic events—the proper preserve of social historians. Collective archiving practices are shifting with digitized and mobile media, however. Leaving to one side the specific archiving projects that seek to digitize material artifacts (and collections) explicitly recognized as of national or global importance,[22] the nature of what we might recognize as a collective archive is shifting with new and mobile media. Relational and interactional aspects of autobiographical narrative change their quality in relation to mediated connections between network Web sites such as Myspace, Bebo or Facebook, participation in which constitutes a cross between a personal Web page or mobile blog, a "networking" space, and a display or performance of autobiography. These kinds of mediated configurations demonstrate the connections between self- and social-, and a redefined *collective archiving*. Newspaper, magazine, and Web site articles are increasingly questioning the public nature of the self-archive—those memories posted on the Web or via mobile connections—when anyone connected to the Internet or via a mobile social network might access some intimate details of participants' lives in ways that directly connect a living, mobile person to information and commentary about them. This redefined collective archiving provides collectively generated accounts of individuals' lives, accounts that are negotiated in the intersections of public and private, in new ways. As Stepinsky notes:

Although both personal and collective memories emerge as products of relationship, it is clear that collective memory acquires a solidity and overarching significance that transcends the individual. Although the operation of individual memory is richly complicated, filled with the possibilities of self-discovery and self-deception . . . collective memory is the ground out of which individual memory, and its particular rhythms, is made possible.[23]

A slightly different configuration of mobile connection to the collective archive is the case of "citizen journalism," where subjective memories of participation in (or witnessing of) public events are mediated by personal mobile technologies, and directly communicated to third-party media agencies for widespread coverage and broadcast. As archival records, however, increasingly they come to inhabit a "public" domain, as what was formerly private and intersubjective becomes increasingly part of collective memory. One example is the phenomenon of "citizen journalism," embodied in the United Kingdom with the July 2005 bombings in London, where those involved posted mobile images of the event to news pages immediately upon receipt of a cell signal. This is a case of the self being positioned within the collective archive of a wider social group or society, rather than the collective archive being itself comprised of self-published memories. Both cases demonstrate, however, that the terms through which we define what constitutes collective memory are changing. In modernity the collective archive was framed in terms of "a set of *social representations concerning the past* which each group produces, institutionalizes, guards and transmits through the interaction of its members,"[24] securing social identities and maintaining ways of seeing and understanding shared histories. Increasingly, however, a new collective memory is simultaneously a set of practices that "remember" the present (and perhaps the future) as the narrations of lives become public while mobile and moving through the world, and moments are captured and collectively circulated in "real time." As Hoskins notes with respect to real-time televisual media, when mobility shifts the terms of collective memory-making from what should be remembered to what could be remembered, the "new memory" to emerge is "a somewhat shaky and schizophrenic televisual past only recognizable at times through flashframes that are repeated rather than 'remembered.' "[25]

While collective archiving practices make explicit some of the shifts in personal and sociocultural memory-making prompted by mobile-media technologies, the new memory of the collective archive further intersects in uneasy ways with those practices that are not only collective but explicitly institutional in nature, organizational in origin, and increasingly automated. This is the memory of the state, of the large-scale organization, of the school, the hospital, the economic enterprise. This is the locale where the diverse practices of collective memory-making via mobile media intersect with the increasingly large-scale and ubiquitous data-gathering activities associated with "information societies"—where the identities generated via autobiographical, interpersonal, and collective narratives of movement through life become enmeshed (albeit unevenly) with the ways of seeing, knowing, and classifying enabled by large-scale databases. As Caroline Bassett notes,

> Through contemporary processes of surveillance of all kinds the materials of everyday life become the matter of private records. Citizens and consumers have fewer secrets; as our private lives are hollowed out, our database records, the secret bodies of material that increasingly define us, to which we have no secure access, grow in volume and density. These records are versions of our life stories and may be narrated back to us with some force. And it is these life histories, rather than our immediate actions, that are generally at issue when we enter surveillance systems, which are increasingly designed to secure identities, by reconciling the body with the history.[26]

While there is a trace of institutionalized memory in Leonard's story—he carries with him the hospital files documenting his illness—it is here that Leonard's mobile memory practices diverge from many of our own. For Leonard is not networked, and his story carries with it little representation of the role of media information (as a social infrastructure and institution) in collective memory-making and information gathering. Leonard's story differs from our own to the extent that his attempts to (dynamically and interactionally) reestablish his memories and identity take place in mobile contexts displaced from the ways our own identities are increasingly embedded in and "fixed" by narratives of life and history—institutional memories—which are not of our own making. While personal and collective memories emerge dynamically from constantly interwoven interactions and accounts amongst members of a community, the "values and forms of life"[27]—the classifications and ways of knowing (or remembering)—from which they emerge increasingly assume a pervasive, ubiquitous (and mobile) visibility, generated by systems of data gathering. Furthermore, the ways of knowing and remembering that underpin mobile and ubiquitous visibility are also implicated in the commodification of life stories and histories— commodity memories—changing the cultural value of mediated memory and memory-making.

COMMODITY MEMORIES AND EXPERIENCE ECONOMIES

Mobile media, like iterations of other new media, significantly expand the possibilities for the production of digital artifacts that inform processes of remembering at different social scales. If contested narratives of everyday life can be digitally recorded on the move, minute by minute (whether in active production by individuals or the automated production of digital monitoring systems), the artifacts that do autobiographical, collective, and institutional memory work also proliferate exponentially. On the one hand, the intensified production of digital artifacts opens up possibilities for the creativity and control of individuals, to create media on the move and to remember and memorialize the self, to the self and others, in novel and interesting ways. On the other hand, media may come in increasingly standardized packages, and the scale and reach of automated records of the self and the collective—whether these are mobile phone traffic or moblog data, RFID signals—narrate histories beyond the control of individuals. Both cases provide an opportunity to reconsider questions about how our society is remembering via new mobile-media technologies, who is remembering, what is remembered, and what might be forgotten in the process.

One point worth considering is the social and economic *value* of mobile-media artifacts in contemporary memory-making practices, and the implications of that value for what is remembered and forgotten in the collective construction of identities. The assumption and prevalence of ubiquitous visibility, described earlier, tends to suggest that the value attributed to mobile-media artifacts is situated in a cultural logic that extends far beyond the social value of memory-making practices for an individual and their intimates. Rather, the value of mobile-media artifacts for "remembering" seems to rely on the articulation of the performative narrative of the "self-as-experiencing" with consumer economies that produce digital artifacts as commodities.

One the one hand, the artifacts produced by mobile media are assumed to be an authentic record of an experiencing self, a record that historicizes and memorializes "experience." Under conditions of ubiquitous visibility, however, these artifacts are produced to be consumed, and to be consumed by a collective much wider than a circle of intimates or communities that interpersonally construct identities through memory work. Rather, such a logic commodifies "experience" within a much larger political economy of digital artifacts where the products of mobile media of various kinds may be bought and sold. As is the case Ken Hillis identifies with respect to eBay,

> [t]he dynamic whereby individuals confer credence on the possibility that a trace of the real inheres in and circulates through an online image runs parallel to an accelerating process of self-commodification whereby one can only become a "truly" successful commodity after one is fully willing to see oneself as an image.[28]

The ubiquitous visibility of experience in this logic constitutes what Mark Andrejevic has termed "the work of being watched."[29] As Bassett points out in an expansion of this argument, the production and consumption of digital artifacts that do memory work is therefore:

a particular form of work, valuable or productive partly to the degree to which it is personal or private consumption . . . personal data become an asset . . . a process that accelerates along with the intensification of media systems and their further encroachment into previously unmediated areas of everyday life.[30]

Self-disclosure, as recorded and distributed via (mobile) media, underpins a productive value in a self-commodity system, and memory-making work becomes part of that system. It is precisely the point where abstracted information from dataveillance becomes reattached to the embodied individual as an element of his or her "being in the world" in ever more intimate ways that it intervenes in the processes of collective and institutional memory-making.[31]

Furthermore, the proliferation of routinized information gathering and storage by large organizations produces what Packard calls "commodity memory" via "dataveillance":

> Electronic surveillance also produces commodity memory. For example, Internet cookies, video surveillance, dataveillance, and radio frequency tagging devices generate electronic memory records in real time, the extent of which seemed inconceivable only a generation ago. Individual consumers benefit from these developments when they download enormous amounts of visual, musical, and text-based data into personal computers. Inexpensive recording devices provide affordable "captured memory" that can be used in ways as various as photos in cell phones, court evidence, or homemade Internet child pornography. Simultaneously, private dataveillance companies collect information about individual consumers (often without consent). Dataveillance product is, among other things, electronically captured memory product of past actions or statistics about past actions and conditions.[32]

What is as yet unclear is how the subjective, interpersonal, and collective digital accounts of personhood, identity, and community (and their simultaneous location in experience economies and commodified memories) will meet with the transactional, abstracted, and automated accounts of our past lives generated by routine state and consumer monitoring. What does seem clear is that the (albeit uneven and disaggregated) practices of memory-making via mobile media will be deeply implicated in these processes via their expansion and extension.

Given the tendencies, what are the implications for the production of mobile, mediated memories? It seems tempting to conclude on this basis that "[t]he extraordinary extension of the social capacity of memory mediated by technology is one of the most evident aspects of the typically modern contradiction between the exponential growth of 'objective culture' and the relative atrophy of 'subjective culture' described by Georg Simmel at the beginning of the century."[33] The social and cultural research currently in its infancy is likely to suggest a somewhat more complex and diverse set of practices and processes, however. Certainly, the mechanisms by which "mobile memories" are increasingly produced, circulated, and consumed are changing the "scapes" (Appardurai) by which the mediation of memories and identities are constructed. Like other media before them, mobile media shift the terms of technological practice, subjective experiences, ideological commitments, cultural logics, and institutional arrangements of mediated memories.

It is worth remembering, however, that memories and the identities that attach are always hybrid, performed, narrated, negotiated, and contested across overlapping social scales. And despite institutionalized attempts to create a seamless circuit of monitoring where abstracted data are reattached to living persons alongside their routine self-disclosure, dataveillance remains imperfect (whatever its sometimes serious and disturbing repercussions of various sorts, for individuals). Machines, too, may forget. Moreover, while mobile media may be complicit in the productive attempts to harness life-experience-as-work in the present, memory-making is far more uneven than this when addressed at different orders of magnitude in practice. If narratives of the past construct and perform identities in various ways, it is tempting to argue that current institutionalized, cultural, and mediated conditions challenge or destabilize the value of past remembrance in Huyssen's "cultural amnesia." As Jedlowski notes, however, the problem is not so much that the past is erased[34] but that the proliferation of (mobile) media present infinitely more possibilities both for the production of memories of various kinds (including both the past and life-in-the-present, as neither can be separated from the other) and also infinitely more possibilities for their selection and forgetting. Mobile media certainly alter the terms upon which autobiographical, interpersonal, and collective memory are organized, but this may be construed as productive and transformative as much as destructive and stultifying.

The increasingly complex materiality of memory-making practices in mobile media, the diversity of networks and artifacts that both produce and are produced by technosocial mobilities, give rise to connections that are distributed, circulatory, irregular, and unequal—both visibilities and invisibilities in social memory. It is also worth remembering that the constructions of memory are not simply ideological commitments or positions in cultural logic. Memory-making is also embodied practice—not only in the specific bodily practices that manipulate technological artifacts to produce digital content (or the manipulation of that content) but social and ritual practices that enact the past in the present (and vice versa)[35] via digital, yet embodied, mediation. As such, it is perhaps in these digital/embodied spaces that "the struggle of memory against forgetting" lies in an increasingly mobile mediated culture.

NOTES

1. Milan Kundera, *The Book of Laughter and Forgetting* (London: Penguin, 1983), 3.
2. John Grech, "Living with the Dead: Sharkfeed and the Extending Ontologies of New Media," *Space and Culture*, 5(3) (2002): 218.
3. Paolo Jedlowski, "Memory and Sociology: Themes and Issues," *Time and Society*, 10(1) (2001): 29–44.
4. Andreas Huyssen, *Twilight Memories: Marking Time in a Culture of Amnesia* (London: Routledge, 1995).
5. Andreas Huyssen, *Urban Palimpsests and the Politics of Memory* (Stanford, CA: Stanford University Press, 2003); Zygmunt Bauman, *Liquid Modernity* (Cambridge: Polity, 2000).
6. Jose Van Dijck, "From Shoebox to Performative Agent: The Computer as Personal Memory Machine," *New Media and Society*, 7(3) (2005): 312.
7. Noel Packard, "Introduction: Sociology of Memory," *American Behavioral Scientist*, 48(10) (2005): 1295.
8. Christopher Nolan, dir., *Memento*, Newmarket, Summit Entertainment, Team Todd, Pathe, 20th Century Fox, 2000.
9. Jeffrey Stepinsky, "Global Memory and the Rhythm of Life," *American Behavioral Scientist*, 48(10) (2005): 1390–91.
10. Marcel Mauss, "Techniques of the Body," *Economy and Society*, 2(1) (1973): 70–88; Michel Foucault, *Discipline and Punish* (New York: Vintage, 1995); Rafael F. Narvaez, "Embodiment, Collective Memory and Time," *Body and Society*, 12(3), (2006): 51–73.
11. Paul Connerton, *How Societies Remember* (Cambridge, Cambridge University Press, 1989).
12. Jose Van Dijck, "From Shoebox to Performative Agent," 323–24.
13. Jedlowski, "Memory and Sociology," 29–30.
14. McKenzie Wark, *Virtual Geography: Living with Global Media Events* (Bloomington: Indiana University Press, 1994); Jean Baudrillard, *The Gulf War Did Not Take Place* (Bloomington: Indiana University Press, 1995); Paul Virilio, *Speed and Politics: An Essay on Dromology* (New York: Semiotext(e), 1977 [1986]).
15. Andreas Huyssen, *Twilight Memories*, 253.
16. Andrew Hoskins, "Television and the Collapse of Memory," *Time and Society*, 13(1) (2004): 118.
17. Stepinsky, "Global Memory and the Rhythm of Life," 1395.
18. Celia Lury, *Prosthetic Culture: Photography, Memory and Identity* (London: Routledge, 1998), 184; cited in Hoskins, "Television and the Collapse of Memory," 118.
19. Hoskins, "Television and the Collapse of Memory," 124.
20. Grech, "Living with the Dead," 216–17.
21. Hoskins, "Television and the Collapse of Memory," 124.
22. See Stepinsky, "Global Memory and the Rhythm of Life," for accounts of, for example, UNESCO Memory of the World Program.
23. Stepinsky, "Global Memory and the Rhythm of Life," 1386.
24. Jedlowski, "Memory and Sociology," 33.
25. Hoskins, "Television and the Collapse of Memory," 124.
26. Caroline Bassett, "Forms of Reconciliation: On Contemporary Surveillance," *Cultural Studies*, 21 (2007): 82.
27. Stepinsky, "Global Memory and the Rhythm of Life," 1386.

28. Ken Hillis, "A Space for the Trace: Memorable eBay and Narrative Effect," *Space and Culture*, 9(2) (2006): 153.
29. Mark Andrejevic, "The Work of Being Watched: Interactive Media and the Exploitation of Self-Disclosure," *Critical Studies in Media Communication*, 19(2) (2002): 232.
30. Bassett, "Forms of Reconciliation," 90.
31. Bassett, "Forms of Reconciliation," 90.
32. Packard, "Sociology of Memory," 1295.
33. Jedlowski, "Memory and Sociology," 38.
34. Jedlowski, 39–40.
35. Connerton, *How Societies Remember.*

Contributors

Gabriele Balbi took a BA in communication sciences and an MA in multimedia and mass communication at Turin University. Balbi has been a PhD student in communication sciences at the University of Lugano since September 2004 (thesis on the early history of Italian telephone). In 2007, he was Visiting Fellow at Harvard University and Visiting Research Student in Maastricht University. Balbi's research interests are related to media history; social history of communication technology (telephone, cell phones); gas and electric light; early broadcasting systems; history of radio and television in Italy and in Switzerland; literature as a source for media history.

Michael Bittman is professor of sociology in the School of Behavioural, Cognitive and Social Sciences at the University of New England. He has published on a wide variety of topics to do with social change and work/family balance in Australian, British, and American journals. His most recent book (coedited with Nancy Folbre) is *Family Time: The Social Organization of Care* (2004, Routledge).

Wendy Van den Broeck is a researcher at IBBT-SMIT, Vrije universiteit Brussel (VUB). Her main expertise and research focus is the domestication of innovative media technologies in a home context. Her current research focuses on TV-related technologies.

Jude Brown is a research fellow in the discipline of sociology, School of Behavioural, Cognitive and Social Sciences at the University of New England. She is a specialist in applied social statistics, has a strong interest in methodological innovation, and has coauthored articles and reports on social disadvantage and the time-use of children.

Axel Bruns is the author of *Blogs, Wikipedia, Second Life, and Beyond: From Production to Produsage* (New York: Peter Lang, 2008). He is a Senior Lecturer in the Creative Industries Faculty at Queensland University of Technology in Brisbane, Australia, and has also authored

Gatewatching: Collaborative Online News Production (New York: Peter Lang, 2005) and edited *Uses of Blogs* with Joanne Jacobs (New York: Peter Lang, 2006).

Liu Cheng (frogcheng99@yahoo.com) is a senior engineer and the vice-director of the information and network center of Yunnan Daily Press Group, Yunnan, China, with a special interest in Internet/online information services. He has the responsibility for the management of network/information systems and the Web site (www.yndaily.com), as well as the promotion and application of new media in Yunnan Daily Press Group. He was a visiting scholar in the Creative Industries Faculty at Queensland University of Technology from February 2006 to February 2007.

Jan Chipchase is one of a team of researchers and anthropologists working at Nokia. Based within the design organization at Nokia, his job is to study people around the world—how they behave, communicate, and interact with each other and the things around them. He shares his observations and insights with Nokia designers, who often accompany him on field trips, helping them to create new ideas for how mobile devices will look, work, and will be used in the future. For more information about Chipchase's work and to catch up on his latest research you can visit his blog at http://www.janchipchase.com/.

Kathy Cleland is a curator, writer, and lecturer specializing in new media art and digital culture. She lectures in the Digital Cultures Program at the University of Sydney (http://www.arts.usyd.edu.au/digitalcultures) and is currently completing a PhD investigating avatars, digital portraiture, virtual characters, and representations of the self in virtual environments.

Kate Crawford is an associate professor in the Journalism and Media Research Centre at the University of New South Wales. She is the author of *Adult Themes: Rewriting the Rules of Adulthood* (Macmillan), which won the individual category of the Manning Clark National Cultural Award. Kate is conducting a three-year study with Professor Gerard Goggin into the uses of mobile media across Australia. She is on the management committee of the Cultural Research Network and is a board member of *Media International Australia* and the *Fibreculture Journal*.

Stuart Cunningham is director of the Australian Research Council Centre of Excellence for Creative Industries and Innovation, Queensland University of Technology. Known for his contributions to media, communications, and cultural studies, he works to promote their relevance to industry practice and government policy. He is the author or editor of several books and major reports, the most recent of which include *The Media and Communications in Australia* (with Graeme Turner, 2006) and *What Price a Creative Economy?* (2006). A collection of essays, *In the Vernacular*, is forthcoming from University of Queensland Press in 2008. Recent publications are available at http://eprints.qut.edu.au/view/person/Cunningham,_Stuart.html.

Jonathan Donner is a researcher in the Technology for Emerging Markets Group at Microsoft Research India, where he studies the social and economic implications of the use of mobile communication technologies in developing countries. Between 2003 and 2005, he was a Postdoctoral Research Fellow at the Earth Institute at Columbia University. Donner earned a doctorate in communication research from Stanford University in 1999, and has worked for Monitor Company and the OTF Group, both consultancies in Boston, Massachusetts.

Leopoldina Fortunati is professor of sociology of communication at the Faculty of Education of the University of Udine. She has conducted much research in the field of gender studies, cultural processes, and communication and information technologies. Fortunati is the author of many books and is the editor with J. Katz and R. Riccini of *Mediating the Human Body: Technology, Communication and Fashion* (2003) and with P. Law and S. Yang of *New Technologies in Global Societies* (2006). Fortunati is associate editor of the journal *The Information Society* and serves as referee for many outstanding journals. She is the cochair with Richard Ling of *The Society for the Social Study of Mobile Communication* (SSSMC). Her works have been published in eleven languages.

Petra Gemeinboeck is an interactive media artist and is assistant professor of digital media at the University of Sydney. Petra's artistic practice crosses the fields of architecture, computer science, new media art, and visual culture. In her works, Petra creates scenarios of encounter in which participants renegotiate social territories, whether they are bodies or city landscapes. Her interactive installations have been exhibited internationally at venues including Ars Electronica, Archilab, MCA Chicago, ICC Tokyo, and Fabrica Brighton. Her works have been featured in magazines such as *RealTime Arts*, *ARCHIS*, and *Computer Graphics World*.

Gerard Goggin is professor of digital communication and deputy director of the Journalism and Media Research Centre, University of New South Wales. His books include *Internationalizing Internet Studies* (with Mark McLelland) (2009), *Mobile Phone Cultures* (2008), *Cell Phone Culture* (2006), *Virtual Nation: The Internet in Australia* (2004) as well as two books coauthored with Christopher Newell—*Disability in Australia* (2005) and *Digital Disability* (2003). Gerard is an ARC Australian Research Fellow, and is editor of the journal *Media International Australia*.

Nicola Green is senior lecturer in the Department of Sociology at the University of Surrey. Her background is interdisciplinary, including media, cultural and gender studies as well as sociology, and she has been involved in UK-based and international research on mobile communications and media for a number of years. She coedited one of the first interdisciplinary collections of research articles investigating mobile technologies—*Wireless World*—and has contributed numerous chapters and articles on mobiles and mobility, addressing such diverse issues as mobile technologies and surveillance, time, space and mobility, mobile technologies and community, and the emergence of a "mobile society."

Leslie Haddon is a part-time Lecturer at Media@LSE, where he teaches a course on Media, Technology and Everyday Life and is currently conducting research for the EU Kids Online Project. He is also research associate at the Oxford Internet Institute, a visiting research associate at Chimera (University of Essex). Over the last twenty years he has worked chiefly on the social shaping and consumption of information and communication technologies, covering the topics of computers, games, telecoms, telework, intelligent homes, cable TV, mobile telephony, and Internet use. In addition to numerous journal publications and book chapters, Haddon was coauthor of *The Shape of Things to Consume: Bringing Information Technology into the Home* (with A. Cawson and I. Miles, Avebury, 1995), author of *Information and Communication Technologies in Everyday Life: A Concise Introduction and Research Guide* (Berg, 2004), and main editor of *Everyday Innovators, Researching the Role of Users in Shaping ICTs* (Springer, 2005).

Larissa Hjorth is an artist and lecturer in the Games and Digital Art Programs at RMIT University, Melbourne, Australia. Over the last seven years, Hjorth has been researching and publishing on gendered customizing of mobile communication and virtual communities in the Asia-Pacific, published widely on the topic in journals such as *Convergence Journal, Journal of Intercultural Studies, Continuum, ACCESS, Fibreculture,* and *Southern Review.* Hjorth has also been active in research of gaming, especially in the Asia-Pacific. In 2007 she was BK21 research fellow at Yonsei University (South Korea), co-convened the *Interactive Entertainment* (IE) conference with Esther Milne (www.ie.rmit.edu.au), and edited a special issue of *Games and Culture Journal.* Hjorth has a forthcoming book entitled *Mobile Media in the Asia-Pacific* (London/New York: Routledge) from her fieldwork conducted in the region.

Mizuko (Mimi) Ito is a cultural anthropologist of technology use, focusing on children and youth's changing relationships to media and communications. She has been conducting ongoing research on kids' technoculture in Japan and the United States and is coeditor of *Personal, Portable,*

Pedestrian: Mobile Phones in Japanese Life. She is a Research Scientist at the School of Cinematic Arts at the University of Southern California and a visiting associate professor at Keio University in Japan (http:// www.itofisher.com/mito).

Dong-Hoo Lee is associate professor of the Department of Mass Communication at the University of Incheon, Korea. She obtained her PhD degree from the department of culture and communication at New York University. Her English-language publications include articles in the *Fibreculture Journal* and the *Asian Journal of Women's Studies,* and chapters in *East Asian Pop Culture: Analysing the Korean Wave* (2008) and *Feeling Asian Modernities* (2004). She has also edited a book, *Mobile Girls @ Digital Asia,* in Korean (2006). Her research interests are digital mobile-media culture, transnational media flows in East Asia, and media ecology.

Bram Lievens is a researcher at IBBT-SMIT whose main expertise is on the domestication of new technologies. Currently he is working on different projects regarding new and emerging technologies, services, and applications, mainly within a mobile environment. Bram Lievens is also active within iLab.o, where he performs living-lab research for open innovation in ICT.

Rich Ling is a sociologist at Telenor's research institute in Norway. He has also been the Pohs visiting professor at the University of Michigan. He is the author of the book *New Tech, New Ties: How Mobile Communication Is Reshaping Social Cohesion* (MIT). Ling is also the author of a book on the social consequences of mobile telephony entitled *The Mobile Connection* (Morgan Kaufmann), editor of *The Mobile Communication Research Series,* and an associate editor for *The Information Society.* He received his PhD from the University of Colorado in his native United States.

Misa Matsuda is a professor at Chuo University. Her research focuses on media and society, particularly the impact of new technologies on everyday life. She also specializes in gender and communication. Having researched uses of the telephone since her graduate years, Matsuda is one of the premier commentators on mobile phones and society in Japan, with her work appearing in both academic and popular venues. Her publications about mobile communication include two coedited collections: *Keitaigaku Nyumon* (Yuhikaku, 2002) and *Keitaidenwa to Shakaiseikatsu* (Shibundo, 2001) in Japanese and a coedited collection: *Personal, Portable, Pedestrian* (MIT Press: 2005) in English.

Daisuke Okabe is a cognitive psychologist specializing in situated learning theory. His focus is interactional studies of learning and education in relation to new media technologies. He works as a research associate at Keio University in Japan.

Virpi Oksman is a researcher scientist working for VTT Information Technology in Tampere, Finland. Her research expertise covers media and mobile usability. Her current research focuses on mobile TV and other mobile-media applications. She has published widely in the field of mobile communication.

Jo Pierson is senior researcher at IBBT-SMIT, involved in innovation strategic research on the use of media technologies at home, at work, and in public settings. He lectures bachelor and master courses on socioeconomic issues of the information society and on qualitative research methods at the Vrije Universiteit Brussel (VUB) in the Department of Communication Studies (Faculty of Arts and Philosophy).

Jason Potts is principal research fellow at the CCi at QUT and also senior lecturer in the School of Economics at the University of Queensland. His work is based on the theory and analysis of economic evolution and complexity economics. He has written two books on this topic: *The New Evolutionary Microeconomics* (2000, Edward Elgar), and *The General Theory of Economic Evolution* with Kurt Dopfer (2008, Routledge). His current work focuses on the application of evolutionary economics to the role of the creative industries in economic growth.

Benedetta Prario is teaching assistant and research fellow at the Faculty of Communication Sciences at the University of Lugano (Switzerland). In the years 2005–2006 she has been visiting assistant at the University of Stirling and at the University Bocconi of Milan. Prario is the author of *Le trasformazioni dell'impresa televisiva verso l'era digitale*, published by Peter Lang (2005), and the author of the chapter "Mobile TV and IPTV" in *Media Literacy*. Prario has published scientific articles in international journals, too, and has presented papers at various international conferences. Her research focuses on media strategy, particularly on digital television, IPTV, and mobile TV.

Harmeet Sawhney is professor in the Department of Telecommunications at Indiana University, Bloomington. His research interests focus on how telecommunications networks are envisioned and created. His research articles appear in *Telecommunications Policy; Journal of Broadcasting & Electronic Media; Media, Culture, & Society; Info; Entrepreneurship & Regional Development; Communication Monographs, The Information Society,* and book chapters in edited volumes. He has been visiting faculty at University of Michigan, London School of Economics, and Stanhope Center for Communications

Policy Research, London. He is currently serving as the editor in chief of *The Information Society.*

Aico Shimizu is an ethnographer and visiting researcher at Keio University, Japan. She received her MA in media and governance from Keio University, studied on mobile-phone picture-sharing practice among community members. She has spent the past three to four years thinking about new mobile/media technology use and youth culture in Japan by observing, talking to people in the context of everyday human life and practices. Her current interests focus on bringing these daily emerging culture and technology use practices into future user experience and urban interaction design.

Atau Tanaka bridges the fields of media art, experimental music, and research. He worked at IRCAM, was artistic ambassador for Apple France, has been researcher at Sony Computer Science Laboratory Paris, and was an artistic co-director of STEIM in Amsterdam. Atau creates sensor-based musical instruments for performance, and is known for his work with biosignal interfaces. He seeks to harness collective musical creativity in mobile environments, seeking out the continued place of the artist in democratized digital forms. His work has been presented at Ars Electronica, SFMOMA, Eyebeam, V2, ICC, and ZKM, and he has been mentor at NESTA.

Jane Vincent is a visiting fellow with the Digital World Research Centre (www.dwrc.surrey.ac.uk) and a doctoral candidate with the Faculty of Arts and Human Sciences at the University of Surrey. Vincent researches the social practices of mobile-communications users, in particular children's use of mobile phones, on which she has completed two studies for the DWRC, and older people's emotional attachment to mobile phones, which is the topic of her doctoral thesis. Vincent joined DWRC in 2001 after over twenty years in the European mobile-communications industry. She regularly publishes in journals and books and presents at international telecommunications industry events.

Judy Wajcman is professor of sociology in the Research School of Social Sciences, Australian National University, and a visiting professor at the Oxford Internet Institute. Recent books include *TechnoFeminism* (2004, Polity Press) and *The Politics of Working Life* (coauthored with Paul Edwards 2005, OUP). She is also a coeditor of *The Handbook of Science and Technology Studies* (2008, MIT Press).

Index

CPSIA information can be obtained at www.ICGtesting.com
Printed in the USA
BVOW010646281011

274699BV00005B/31/P